CO

SANTA FE PUBLIC LIBRARY

DISCARD

STILL WATERS

ALSO BY CURT STAGER

Your Atomic Self (2014)

Deep Future (2011)

Our Future Earth (2011)

Field Notes from the Northern Forest (1998)

STILL WATERS

THE SECRET WORLD OF LAKES

Curt Stager

W. W. NORTON & COMPANY
Independent Publishers Since 1923 | New York | London

Copyright © 2018 by Curt Stager

All rights reserved
Printed in the United States of America
First Edition

For information about permission to reproduce selections from this book,
write to Permissions, W. W. Norton & Company, Inc.,
500 Fifth Avenue, New York, NY 10110

For information about special discounts for bulk purchases, please contact
W. W. Norton Special Sales at specialsales@wwnorton.com or 800-233-4830

Manufacturing by LSC Communications
Book design by Brooke Koven
Production manager: Anna Oler

Library of Congress Cataloging-in-Publication Data

Names: Stager, Curt, author.
Title: Still waters : the secret world of lakes / Curt Stager.
Description: First edition. | New York, NY : W.W. Norton & Company, 2018. |
Includes bibliographical references.
Identifiers: LCCN 2017060910 | ISBN 9780393292169 (hardcover)
Subjects: LCSH: Lakes. | Lake ecology. | Lake conservation. |
Nature—Effect of human beings on
Classification: LCC GB1603.2 .S7 2018 | DDC 551.48/2—dc23
LC record available at https://lccn.loc.gov/2017060910

W. W. Norton & Company, Inc.,
500 Fifth Avenue, New York, N.Y. 10110
www.wwnorton.com

W. W. Norton & Company Ltd.,
15 Carlisle Street, London W1D 3BS

1 2 3 4 5 6 7 8 9 0

FOR DAN, JOE, AND DAD

CONTENTS

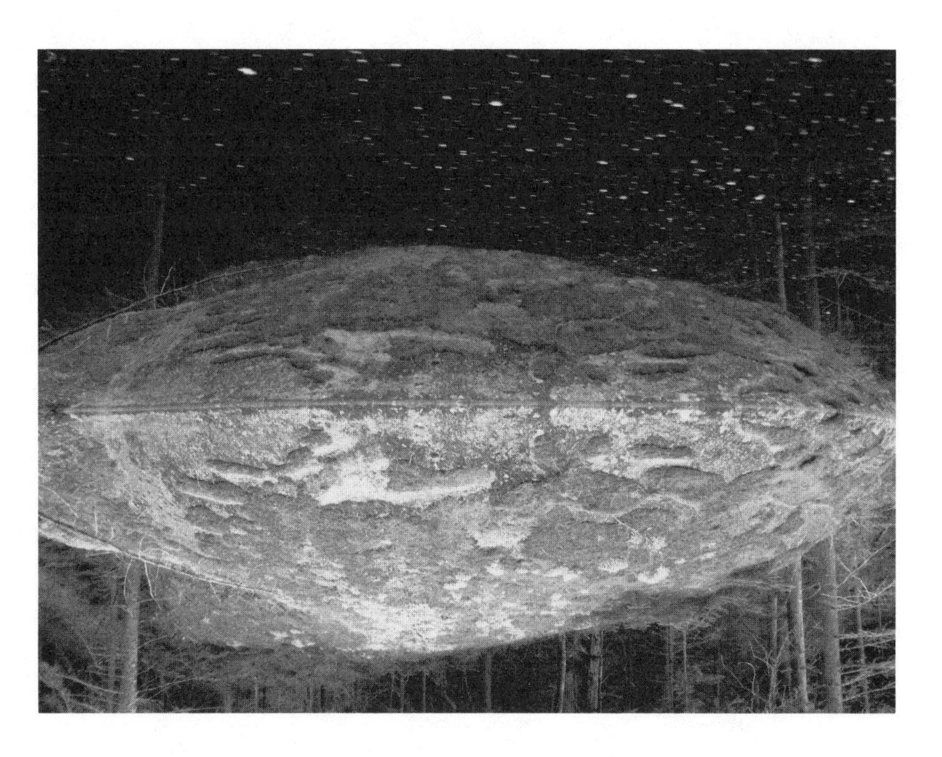

What is life without stars?
And trees to see them through?
And a lake for a cool, clear view
when you need a reflection
to flip perspectives,
and the world?

—HANNA CROMIE

PROLOGUE

STILL WATERS

I come down to the water to cool my eyes. But everywhere I look I see fire; that which isn't flint is tinder, and the whole world sparks and flames.

—ANNIE DILLARD, *Pilgrim at Tinker Creek*

COME DOWN to the edge of the lake. What do you see?

A sparkling blue plain, flat as a field for sailing on? A fishing hole? Perhaps a place where monsters or spirits lurk, a reservoir to water and refresh a city, or a revered symbol of wilderness. It can be all of those things and more, but most of all it is a mirror. What you see in a lake is a reflection of yourself and a glimpse of your relationship to the natural world.

That is what this book is about.

Come to the lake again when the air is restless and take a closer look. Is it really blue? The wind sweeps the surface into dunes that carry the colors of weather, water, and shore, a fluid mosaic that changes faster than you can describe it. Peer into the approaching waves when the Sun shines through them and they may glow green, brown, or multiple shades of blue depending upon the season and the kinds of plankton within them. If you scoop a glass full there may be no apparent color at all, and if you try it in winter the water itself may have turned to glass.

For the most profound experiences, however, come to the lake when the air is calm and the water is still. Now it resembles a hole in the Earth through which to view a subterranean sky, an illusion that also reflects reality. When Henry David Thoreau wrote that "Heaven is under our feet as well as over our heads," he meant it metaphorically, but there really is a sky down there, almost 8,000 miles (12,750 km) away on the far side of the planet. Suddenly, you and the world around you are transformed into something more miraculous and precarious, a mote of life amid a thin plankton of planets within a deep lake of space.

Step even closer, right up to the rim. Now you can see in the shallows that there is a bottom to it, and that the blue plain was yet another mirage. The lake is not a flat sheet but, like you, a body of water atop a landscape whose contours continue beneath it. If your imagination removes the water to reveal that hidden terrain, you will be reminded that much is concealed beneath the surface of daily life. The lake itself takes the shape of a mountain of water that has been clipped, flipped, and planted top-down in the earth. The living community within it also mirrors the layering of mountain habitats, with life least abundant in the least accessible zones where oxygen can be perilously scarce.

Lakes are secret worlds within worlds hiding in plain sight. If you come close enough to sweep your fingers through the water, a fish might dart away at your approach. The fish will then vanish into places beneath the reflections where rich and varied forms of life thrive unseen, especially in the realm of the very small. Creatures of legend take form and lend their names to the nymphs, hydras, and cyclopses who live there.

Now press a wet fingertip against your brow. The coolness of the moist smear is the work of evaporating particles, the primal elements of existence. Early Greeks named them "little lumps of stuff" (*mole-cules*) and "indivisibles" (*a-toms*) without knowing their sizes or properties or even proving their presence. Many of us think of them only as squiggles on blackboards or the stuff of dangerous chemicals and bombs, but they link us to all life, the planet, and the deep history of the universe.

That transformation of water into air reveals elemental connections between you and the lake. Every breath you draw invites lake water vapor into your body, and moisture from your lungs melts into the surface of the lake with each humid breath you release. You also run the alchemy backward, turning air into water. One in ten of your tears is metabolic water that your cells crafted from airborne oxygen, some of which may have been made from water molecules by plankton in the lake before you.

On the surface this book is about lakes, but it is more than that. It is also about evolutionary processes that anchor the roots of all life in water. It is about inanimate air, water, and stone becoming an alga, then a shrimp-like copepod, then a trout, then a loon or, perhaps, a person. It is about tracing your ancestral connections to distant aquatic relatives who are too small to see. It is about a tiny dab of tissue morphing into a wriggling tadpole who later becomes a frog. It is also about people maturing as I did in my growth from inquisitive kid to scientist, and as our own species is doing in this remarkable new slice of Earth history, the Anthropocene Epoch, or the Age of Humans.

These are the kinds of experiences I hope to share with you in *Still Waters*, but I also have another agenda in mind.

More than a century and a half have passed since Henry David Thoreau wrote *Walden*, the world's best-known book about lakes. *Walden* is widely quoted for its commentary on the human condition, but it is also noteworthy for its treatment of lakes as three-dimensional worlds unto themselves rather than mere playgrounds, resources, or reflecting pools. To my knowledge no other book has done this so well for a general audience, but I also believe it is time for an upgrade that is more appropriate for our own century. It is a bold claim, so I will explain it briefly.

Textbooks, manuals, and technical papers on the science of lakes abound, and I recommend some of them to you in the references section, but they are not *Walden*. They are generally dry and inaccessible to the average reader, and most do not readily bring their subject matter down to a meaningful personal level. In my opinion, two of the best and most comprehensive resources among them are G. Evelyn Hutchinson's four-volume *Treatise on Limnology* and Robert Wetzel's *Limnology* textbook.

Other more reader-friendly books with lakes in their titles are more about people than lakes, a disappointment to someone who hopes to better understand what a lake is, how it works, what lives in it, or what challenges it faces in the modern world. In *Walden*, Thoreau blended what was then new scientific information with his own aesthetic and philosophical insights, but we have come a long way since he wrote, "Thank God men cannot fly and lay waste the sky as well as the earth."

With an additional century and a half of scientific discoveries to inform us, we can look more deeply into the natural world and our place in it than Thoreau and his contemporaries could. Travel and literature searches are easier today than during the nineteenth century, and we can now more readily uncover new findings from iconic lakes and tie them more firmly to our lives. Thoreau wrote, "One generation abandons the enterprises of another like stranded vessels." Having thereby been given license by Thoreau himself, we need not hesitate to build upon the most accurate and inspiring parts of his writings.

There is nothing like a lake to reflect and reveal the world, and many of the insights that the study of lakes now offers are much needed in the Age of Humans. Modern science demonstrates that we are not only physically continuous with the planet and shaped by its elements and forces, but also a force of nature in our own right. We are now so numerous, our technology is so powerful, and our lives are so interconnected that we are changing the ecology and chemistry of lakes and oceans, altering climates, and forcing entire species into extinction.

As a scientist I am more of a mapmaker than a navigator, and I won't be able to offer complete answers to some of the practical and ethical questions that arise on these pages. Instead, I simply hope to show in clear and enjoyable ways that the science of lakes can provide valuable and interesting perspectives on the world and our place in it.

We carry many definitions of "Nature" within us, for example, and some of them will be challenged by facts that emerge here. Many of us also cherish a personal or traditional myth of Eden that we refer to in deciding how to use, protect, or restore the lands and waters of the Earth. One chapter will explore the environmental histories of sacred lakes of the Jordan Valley in search of scientific insights against which to measure such worldviews. And although science offers our clearest perspectives on reality, scientists themselves are as fallibly human as anyone else, a point I will illustrate with myself as the prime example in the chapter on sky water.

The applications of science that I prefer to explore here are those that help us to appreciate and listen more carefully to the world rather than imposing our desires and imaginations on it. The language I use in *Still Waters* to refer to animals is consistent with that preferred approach. Rather than deny selfhood to them, I refer to them as conscious beings whenever possible. You will not be told that "*The* Bass guards *its* nest" as though bass were cartoonish robots or there is only one bass in the world. Instead I will say "Bass guard their nests" or "*He* is guarding *his* nest," using the proper gender as well when possible (yes, a male largemouth bass does guard his nest after his mate deposits thousands of tiny eggs there). As we shall see, individual selfhood is the raw material of evolution, a frequently overlooked but important factor in ecology and resource management that is easily observed by anyone who pays close attention to living beings. Biologist Robin Kimmerer is an eloquent champion of this approach to language, and she explains it well in her wonderful book, *Braiding Sweetgrass.* I have used her motto from a recent public presentation, "No itting," as a guide in my own writing.

Trying to absorb the rush of new scientific information can be like trying to drink from a waterfall, and *Still Waters* is not meant to be a comprehensive encyclopedia but a tasty sample to whet your appetite for more. After having explored related topics of geologic time and our elemental nature in *Deep Future* and *Your Atomic Self,* I have applied those unconventional perspectives to selections from the technical literature in this book along with tales of my experiences with lakes around the world during the last six decades.

The following chapters will lead you from a time capsule of American history at Walden Pond to a lesson in natural selection by way of a crocodile attack in Zambia. The atomic elements of the Earth will link you to fish and lake water while you investigate purported miracles at the Sea of Galilee. You will hunt for Scotland's Loch Ness monster from orbiting spacecraft, hike the shore of the world's deepest lake in Siberia, seek the most pristine waters of the Adirondack Mountains in upstate New York, and dive into a tropical crater lake in Cameroon. You will also survey our massive impacts on lakes worldwide from the comfort and safety of a laptop. Along the way, you will learn of scientists who worry that a huge African lake has become too clouded with algae, anglers who are concerned that a lake in Germany has become too clean, resource managers who poison wilderness lakes with pesticides in order to save them, and environmentalists who try to protect such lakes from being poisoned. Perhaps most surprising of all will be the richness of untamed life in unexpected places such as the tiny backyard pond that first introduced me to aquatic wilderness during my childhood in suburban Connecticut.

Thoreau's contemporary Herman Melville also appreciated the powerful connections between water and human beings. In the opening chapter of *Moby-Dick* he wrote, "Let the most absent-minded of men be plunged in his deepest reveries—stand that man on his legs, set his feet a-going, and he will infallibly lead you to water, if water there be in all that region. . . . Yes, as everyone knows, meditation and water are wedded for ever."

May these explorations and meditations help you to see yourself and the world we share in a more miraculous light and inspire you to look beneath the surface of things in ways that our forebears never could, while also savoring the reflections.

Come on in, the water's fine.

(photo by Curt Stager)

STILL WATERS

1

WALDEN

A lake is a landscape's most beautiful and expressive feature. It is Earth's eye; looking into which the beholder measures the depth of his own nature.

—HENRY DAVID THOREAU, *Walden*

MY FIRST VISIT to Walden Pond comes on a quiet morning in May 2015, when the spring air is fresh and fragrant with white pine resin and seasoned with the musty scent of last year's dried oak leaves. I walk slowly to the edge, hunch down, and brush my fingers through the cool, clear water. Ripples roll away from my fingertips and briefly warp the mirrored sky. A few feet ahead of me where the reflections don't hide the bottom, clean sand and pebbles fade into translucent blue mists where sunfish and minnows cruise. In my mind's eye, I also see great depths of time beneath the water itself, recorded in soft layers of sediment. To take a deep dive into history at Walden Pond is to explore our place in nature in ways that Henry David Thoreau, the nineteenth-century philosopher-naturalist who made it one of the most famous lakes in the world, could only imagine during his two-year stay in the one-room cabin that he built near the shore in 1845.

Technically speaking, Walden Pond is a flow-through kettle lake, a dimple in glacial deposits that formed shortly after the last great ice

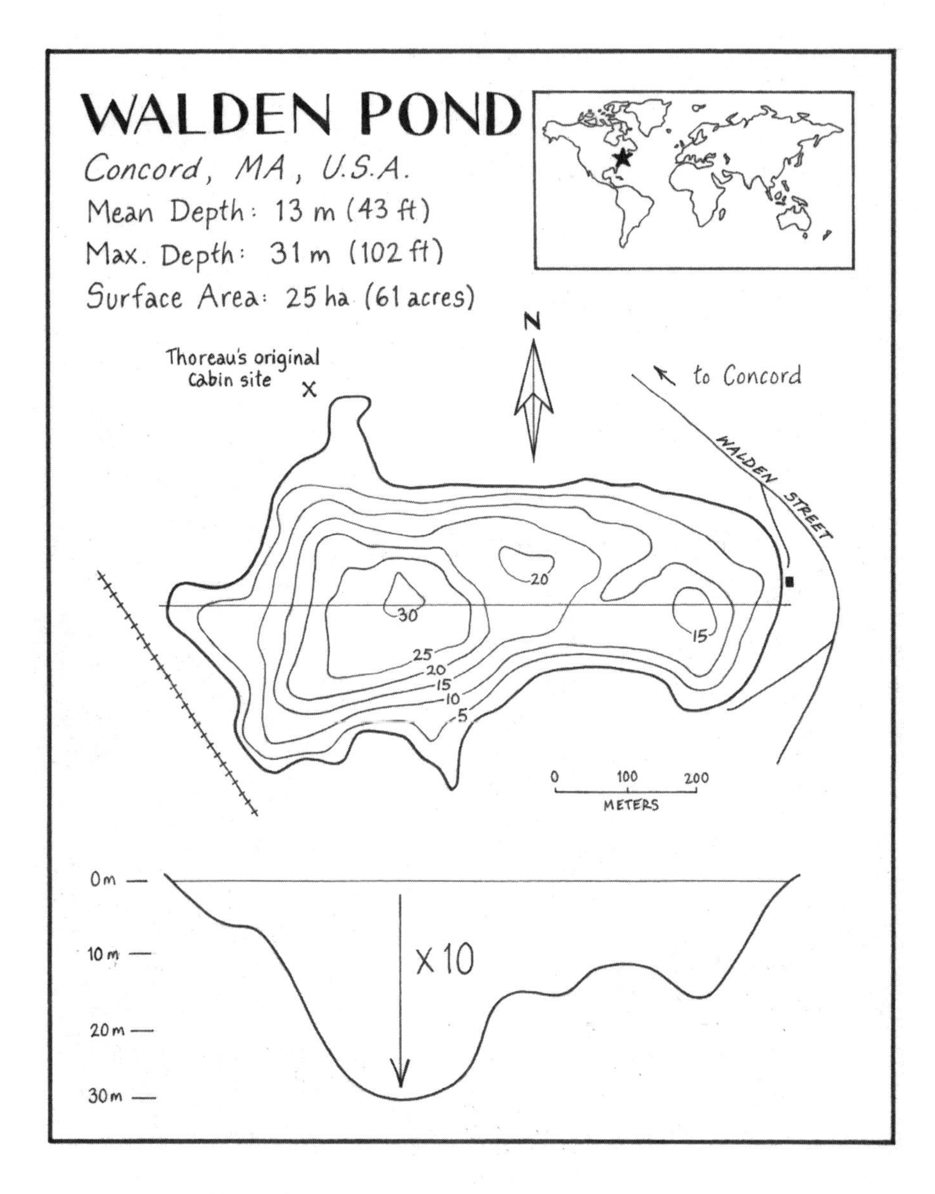

sheet thawed back from eastern Massachusetts roughly fifteen thousand years ago. Meltwater rivers spread sand and gravel amid stranded iceberg islands as one partially buried berg slowly melted down into a gritty pit like ice cream in a cone. Eventually, only the cone remained, exposing the surface of the local groundwater table. As of this spring morning, an extended dry spell has dropped that surface so low that my wife, Kary, and I can walk all the way around it on gently sloping sand rather than using the footpath on the steep, wooded flanks of the kettle.

Walden has what many of us desire in a lake. Gem clarity, weedless shallows that practically beg you to wade in them, fine fishing, a comfortably modest size of 61 acres (25 ha) that limits the swell of waves on a windy day, easy access to a swimming beach, and a launching area for nonmotorized watercraft. The lake is surprisingly deep, 102 feet (31 m) give or take a few depending on the elevation of the water table, which makes it the deepest water body in the state. But it is of course best known as a memorial to Thoreau and his love of nature that draws hundreds of thousands of visitors every year and is cherished by millions more who will never see it.

Because of Thoreau's writings, Walden Pond has become a symbol of wild nature unspoiled by the hand of humanity. It can therefore seem

Thoreau's cove, Walden Pond. *(photo by Curt Stager)*

strange to visit it in the company of sunbathers, joggers, and busloads of foreign tourists, who are often present. The Massachusetts Department of Conservation and Recreation maintains a public beach and bathhouse on the eastern shore that can host more than half a million people over the course of a summer. This early on a spring morning, though, Kary and I have the place mostly to ourselves.

Hidden in the forest at the head of a cove on the northwestern shore is the former site of Thoreau's cabin, which he built on the property of his friend and mentor, Ralph Waldo Emerson. The cabin is long gone, but the site of its foundation is marked off with posts. Beside it is a low mound of cobbles about 10 feet (3 m) across that changes in shape and structure as pilgrims bring new contributions to it from all over the world. Many of the stones are inscribed with magic marker or paint. We kneel down for a closer look as the cove glitters through the trees behind us.

"I want to thank you Thoreau. Just from hearing about you I am enlightened."

"Uly Walden Carr—R.I.P. Love, Dad."

"In nature, we are one."

A careful reading of *Walden; or, Life in the Woods* makes it clear that Thoreau never intended his cabin to be a solitary hermitage, although fans and detractors alike often misunderstand this. It was more an author's workshop than a fortress of isolation, and throughout his lakeside residency he often visited family and friends in Concord and entertained guests at Walden. Ice-cutters and woodcutters, anglers and boaters, and even a noisy train were as much a part of his surroundings as the lake, woods, and wildlife. He retreated to the cabin largely in order to write in a quieter setting than he could find in town and to "live deliberately, to front only the essential facts of life, and see if I could not learn what it had to teach, and not, when I came to die, discover that I had not lived."

Thoreau's cabin experiment was also a field test of the transcendentalist philosophy that Emerson championed. For Emerson, nature represented an embodiment of the divine, an aesthetic ideal that was best described in poetic or quasi-religious abstractions. Contemplating it was a way to transcend normal daily life and seek deeper spiritual lessons. Emerson believed that nature was "all that is separate from us, all which Philosophy distinguishes as the NOT ME," and "essences unchanged by man; space, the air, the river, the leaf." Such ideas still resonate with many of us today. Recognizing that satellites now cruise space and our fossil carbon emissions contaminate the air, rivers, and leaves of the entire planet, author Bill McKibben built a similar concept into the title of his groundbreaking book on global warming, *The End of Nature*.

Thoreau's equally reverent views, however, were more explicitly anchored in physical reality than Emerson's, the product of both aesthetic and scientific sensibilities. His journals recorded minute details of the world around him, from the number of growth rings in a tree stump to the gyrations of shiny black whirligig beetles on the surface of the lake.

I see no whirligigs here this early in the year, but they are easy to spot on a lake such as Walden when the water is still and they can gather in close, swirling clusters. They overwinter on the bottom and emerge in spring to breed, producing new generations that grow to fingernail length within a few weeks. Each beetle uses flattened legs to paddle quickly through the thin surface film, guided by compound eyes that are each divided, with one-half aimed above the water line and one-half below. Most fish leave whirligigs alone because they leak bitter chemicals when handled, and I have seen newly stocked brook trout, brazen and ignorant from life in the hatchery, snatch whirligigs from below and then spit them back out again like slippery watermelon seeds. Whirligigs often gather in groups that help to discourage predators by pooling more watchful eyes in one place, and the whirling dances within the clusters are not as random as they seem. The individuals on the perimeter are generally searching for fallen gnats, emerging midges, or anything else edible, and they emit ripples like radar to home in on struggling prey. In adult swarms, those closer to the center are more likely to be cruising for mates, using their ripples to communicate with one another and avoid collisions.

Much more has been said and written about Thoreau's philosopher-poet side than his naturalist side, but as a scientist I am more interested

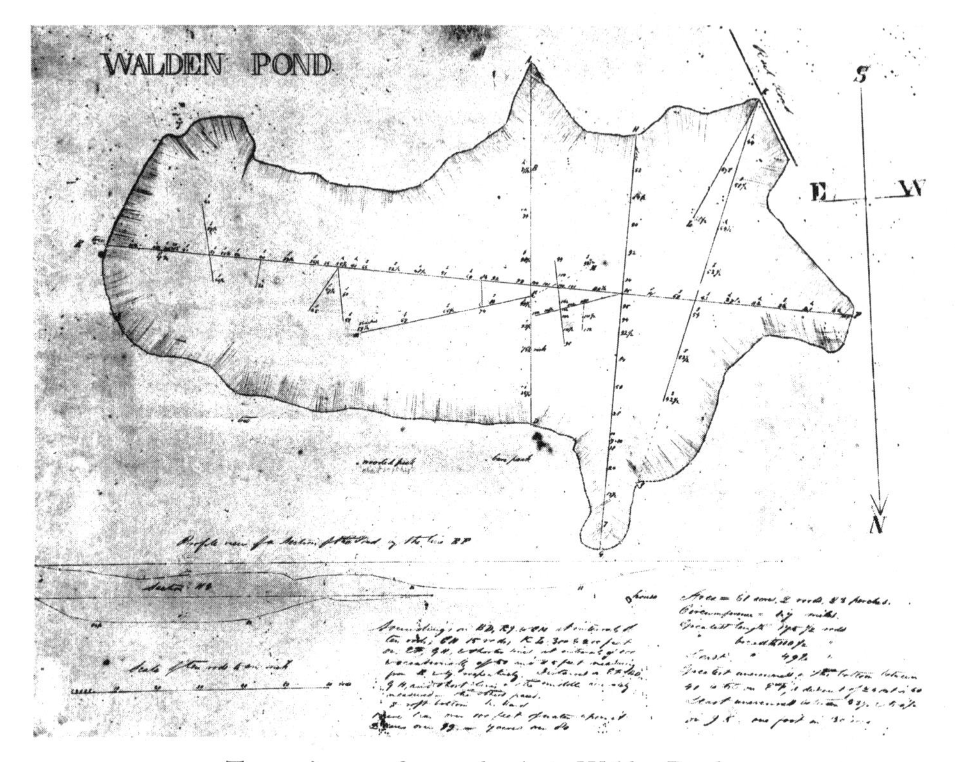

Thoreau's map of water depths in Walden Pond.
(courtesy of the Concord Museum, Concord, MA)

in the latter. The journals that he kept from 1837 to 1861 were so full of natural history observations that they might have become a major scientific work if he had not died of a lung ailment at age forty-four. He probably thought so, too. Two months before his death in 1862 he wrote a letter to a friend, saying, "if I were to live, I should have much to report on Natural History generally."

During the winter of 1846, Thoreau drilled more than a hundred holes through the ice of Walden Pond and lowered a weighted line to produce what may be the first map of the floor of an American lake, thereby identifying Walden's deepest point in the western basin near his cove. In August 1860, he also sent a thermometer down in a stoppered bottle to measure the layered structure of the water column, a first formal analysis of the thermal stratification of the lake. He was amazed at the temperature difference between the upper and lower layers, and he speculated on what it might mean for the resident fish. "What various temperatures, then, the fishes of this pond can enjoy," he wrote. "They can in a few minutes sink to winter or rise to summer. How much this varied temperature must have to do with the distribution of the fishes in it."

In August 1939, lake ecologist Ed Deevey made similar measurements from a rowboat and confirmed Thoreau's reports. He also measured the stratification of the water in more detail, finding temperatures close to 79°F (26°C) in the upper 15 feet (5 m) that fell to 41°F (5°C) near the bottom. Writing in *Quarterly Review of Biology*, Deevey noted that Thoreau's curiosity was "unusually fruitful when directed toward lakes," and called him the first American limnologist, or lake scientist.

Other scientists have also used Thoreau's observations in their own research. The Boston University ecologist Richard Primack has compared recent observations of ice-out dates, flowering times, and other signs of spring to the dates that Thoreau recorded in his journals. In *Walden Warming*, he used those data to show that climate change has shortened the ice-cover season by several weeks since the nineteenth century. And one journal entry from 1854 tripped up another friend of mine, biophysicist Charles McCutchen.

When I visited him recently at his family property on Lake Placid in upstate New York, Charlie was in his nineties but still busying himself in retirement with research projects and inventions, including wooden contraptions that harness wave power to propel themselves and a motor launch that smooths its own bouncing in rough water with a spring-loaded keel. While standing beside a local stream in 1970, Charlie had noticed something resembling a fine thread on the surface that undulated crosswise to the current. After careful study, he identified it as an ephemeral wrinkle where the surface film folded inward on itself. Soon after he published his discovery in *Science*, however, another researcher pointed out that Thoreau had already described the same phenomenon, both accurately and more poetically. "It is interesting," Thoreau wrote, "to distinguish the different surfaces,—here broken into waves and sparkling with light . . . and there quite smooth and stagnant. I see in one place a sharp and distinct line, as if it were a cobweb on the water . . . as if it were a slightly raised seam."

As we watched the sunlight sparkle on Lake Placid, it seemed to me that Charlie relished the thought of being scooped by Henry David Thoreau.

WHEN I RETURN to Walden in August, my students Rory and Elliott carry our two canoes to the boat launch and lash them into a makeshift catamaran. It is easy to see that the lake has changed since Thoreau's time in ways that he and Emerson would probably dislike. The adjacent beach is packed with bathers, and although the water is still clear, a faint greenish tinge warns of potential trouble.

Analyses published in 2001 by the United States Geological Survey showed that surreptitious urine releases by swimmers had approximately doubled the summer phosphorus budget of the lake. Phosphorus, whose elemental symbol is the letter "P," is a key structural atom in cell membranes, energy-storing molecules, and genes, and it is therefore a common currency in the world's food webs. All living things, ourselves included, consume it in food and release it in waste molecules that other organisms may later use. Humanity's new role in the Walden Pond ecosystem as a key source of pee-P for algae is also reflected in the results of sediment core studies that were conducted in 1979 by University of Wisconsin researcher Marjorie Winkler and in 2000 by Canadian ecologist Dorte Köster and colleagues. They found that distinctive phosphorus-loving species have dominated the planktonic algal community since the early twentieth century. My students and I have come here now to consult the sediments for an update on the status of the lake and to more closely examine its climatic history with an eye toward the future.

A middle-aged man with a fishing rod in his hand pauses to ask if I know where the lure-snagging weed beds are. A fibrous alga, *Nitella*, forms a ring of matted meadows on the lake bed at depths between about 20 and 40 feet (6–13 m), but darkness prevents it from coloniz-

Nitella from Walden Pond. *(photo by Curt Stager)*

ing the deeper places farther offshore. Thoreau mentioned it in *Walden*, calling it "a bright green weed (that) is brought up on anchors even in winter." It is common in clear-water lakes and resembles a plant, but it lacks flowers, seeds, or vein-bearing stems and roots. Unlike true aquatic plants, its ancestors remained within an entirely different kingdom of life, Protista, that is dominated by single-celled species. The *Nitella* meadows in Walden Pond divert dissolved phosphorus away from the microscopic algae of the plankton and trap it on the bottom. Like the continuous flush of groundwater, they help to keep the lake clear, but ecologists worry that any further clouding of the water by overfed plankton might shade them out and tip the scales in favor of pond scum.

To this fisherman, however, the *Nitella* seems to be more of a nuisance than a blessing. I ask what he hopes to catch. "Rainbows and browns," he says. Neither species lived here in Thoreau's day. Rainbow trout are native to the western United States, and brown trout were brought to North America from Germany during the nineteenth century. Despite the lake's status as a revered symbol of wilderness, county officials had it poisoned with a pesticide, rotenone, in 1968 in order to remove "trash fish" such as the pout and pickerel that Thoreau once knew and to make way for nonnative game species.

I walk a short distance along the shore and nearly step on another sign of humanity's role in the spread of species around the planet, a softshell turtle who has been sunning herself on the water's edge. Her smooth, flat shell is more than a foot wide, and the speckled tan disc blends well with the sandy background. When I kneel for a closer look, she launches into the depths like a flying saucer. Softshells are rare in New England and not native to Walden. This lone female is said to be a former pet who sometimes scoops a nest in the sand with her hind legs and lays leathery eggs in it, but as she has no mate to share the lake with her, the eggs never hatch.

We know how the softshell turtle, rainbow trout, and brown trout got to Walden, but how did the first native fishes arrive without the aid of surface inlets or outlets? Coming up with answers to this question can be more entertaining than finding the truth of the matter, which remains elusive. Did they colonize the lake when glacial meltwater still flooded the area? Did birds of prey drop them in by accident? A friend of mine once told me that a fish landed alive and flopping at his feet while he was walking on a woodland trail in the Adirondack Mountains, an escapee from an osprey's talons. Or did early indigenous people carry them here in their own paleolithic stocking programs? For now, at least, Walden is keeping that story secret.

A young free-diver approaches and shows me a GoPro video on his cell phone. In the video, his hands follow a guide-rope into murky darkness in the hundred-foot basin. According to him, the deepest part of the lake is surrounded by steep ledges, and huge snapping turtles lumber around the margins of the pit like dinosaurs. "There's no light down there," he explains, "so you can't tell where the bottom is. Sometimes I do a head-plant into the mud because I can't see where I'm going." Fortunately, our coring site will be in an adjacent basin that is—hopefully—less pitted with head-plants.

As we paddle out to the center of the lake a bald eagle swoops low overhead, perhaps scanning for trout. After the huge bird flaps back up and over the tree line, my attention aims downward, too. Beneath us lies an extension of the landscape that mirrors the underbelly of the iceberg that formed it. Thoreau identified the hundred-foot hole at the west end of the lake and a 55-foot (16 m) basin at the east end near the swimming beach, but he missed a third one midway between them. The USGS scientists found it only a decade ago, measuring depths close to 65 feet (20 m) near the center of it. That is where we are headed now.

Rory and Elliott toss two anchors and draw the lines tight while I retrieve a conical net that I have been towing behind us. The mesh is finer than that of a nylon stocking, and it sieves the dilute broth of plankton beneath us. When I hold a glass vial of the catch up to the sky, I see creamy flecks dancing like dust motes in the sunlight. These shrimp-like copepods and cladocerans are the main prey of Walden's fingerling fish and minnows. They use their tapered abdomens as rudders and swim by paddling with multiple pairs of jointed limbs while additional limbs also strain the water for microscopic algae. A healthy population of zooplankton (animal plankton) can filter the entire volume of a lake within days, a testament to the rapid growth of the phytoplankton (plant-like plankton) they graze on. I tip most of these animals back

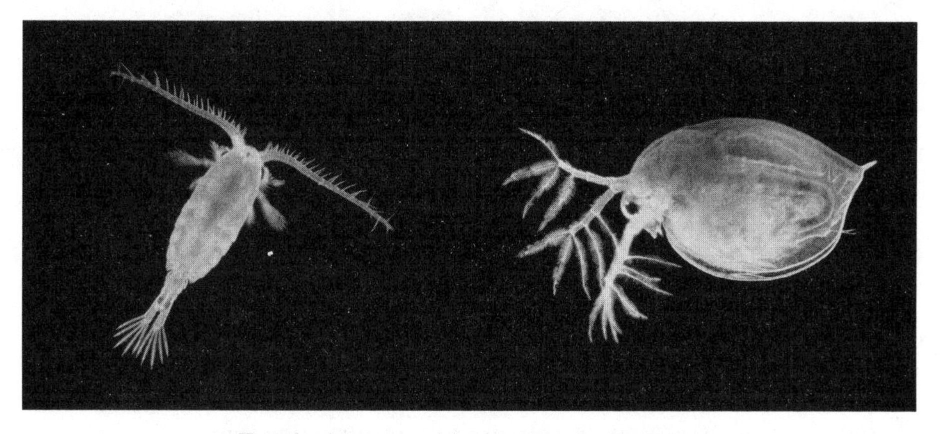

Zooplankton, roughly the size of salt grains.
Left: Copepod. Right: Cladoceran. *(photos by Mark Warren)*

into the lake and feel sorry for the few I must keep as specimens even though I have already killed many of their kind every time I gulped a mouthful of lake water or toweled off after a swim.

All around me, trillions of living specks such as these are feeding, breeding, dying, and ultimately sinking to the bottom. Joining them in the gentle flurry of debris are leaves, twigs, and puffs of pollen from the forest. Mushroom spores, insect wings, and translucent grains of beach sand from the shore. Genes and bones from fish and turtles, and the gleaming glassy shells of microscopic diatom algae. I lean over the gunwale to look through the wavering halo of sunbeams that surround my silhouette, imagining the detritus of life settling like snow beneath me. Each successive layer represents a page in the history of the lake and its surroundings. When our free-diving friend next plants his head in the soft brown ooze of the main basin, his scalp will push through decades of accumulated crud. If he were to plunge his hands to the wrists in it, he could touch plankton that also brushed Thoreau's hands during swims a century and a half ago.

We deploy the smaller of two core samplers first, in order to confirm that we are positioned far enough from the *Nitella* meadows to avoid clogging the core barrel with fibers. The collection tube is made of clear plastic that lets you see at a glance how much mud you've retrieved. A heavy brass cone surrounds part of the barrel to help the sampler sink into the sediments under its own weight. The most distinctive feature is a loop of rubber tubing attached to a white plastic ball that snaps smartly into place over the mouth of the barrel like a protective hand as it emerges from the mud. The whole thing breaks down into suitcase-sized pieces as if it were a collapsible weapon in a secret agent film, and I have carried this snazzy little rig with me on numerous expeditions around the world. My only criticism of it is that its odd appearance occasionally makes a suspicious airport security officer raise an eyebrow and ask, "You use this thing to do . . . *what*?"

Rory and Elliott lower the sampler hand over hand until they feel the line go limp. When it splashes aboard moments later, Rory holds it upright to avoid disturbing the loose, flocculent surface layers. The core is as long as her forearm and resembles a tube of chocolate pud-

Students with a sediment core at Walden Pond. *(photo by Curt Stager)*

ding. No sign of *Nitella* here, so we stow the first sample and lower a longer, heavier, home-built device overboard. This one is equipped with a counterweight that feels the bottom and triggers a release mechanism just before the base of the core barrel meets the mud. Moments later, 33 inches (84 cm) of lake history break the surface.

The results of previous coring studies suggest that this sample represents about 1,500 years, which carbon-14 dating of the mud will later confirm. A band of sediment that was deposited during Thoreau's lifetime lies 8 to 9 inches (20–24 cm) below the surface layer of the core. I can link the two strata with the span of two hands, a distance that will eventually shorten down there on the lake bed as new, watery mud gradually compresses under the weight of future layers. Thoreau's writings easily draw your imagination back to the nineteenth century, but a sediment core such as this pulls you even deeper into the past by encouraging you to ask "what happened before *that*?" Seeing so many relics of Walden's yesteryears stacked one atop the other in this manner exposes self-centered views of history for what they are, reflections of our own minds that obscure our momentary positions in an open-ended river of time.

I GREW UP thinking of myself as an environmentalist, and Emerson's and Thoreau's writings fed my youthful passion for the natural world. Now as a senior scientist, I still love wild places but I have learned over the years that some of my earlier ideas were incompatible with ecological history, and I am no longer sure that lakes are necessarily "unnatural" in the presence of humans.

Walden Pond was already heavily influenced by people during the nineteenth century, and Thoreau himself recognized that other human beings had known the place long before he had. He wrote of finding stone projectile points and bits of pottery near the lake, but his published works revealed few details of the region's deep human history.

Concord was founded in 1635 by colonial settlers who purchased parcels of land in what was then known as Musketaquid from members of the Massachusett Federation, a tribal alliance whose territory stretched from the Connecticut River to what is now Boston. Ironically, descendants of the Massachusett are not officially recognized as a tribe by the state that now bears their name, just one of many symptoms of a widespread cultural blindness to the long presence of human beings in the Americas. Such historical myopia is not new, and a sardonic entry in Thoreau's journal that describes the founding of a historical museum in Concord acknowledged it. "We cut down the few old oaks which witnessed the transfer of the township from the Indian to the white man," he wrote, "and commence our museum with a cartridge box taken from a British soldier in 1775." Attenuated views of New England history were further clouded by the near-extermination of the Massachusett, both physically and culturally, two centuries before Thoreau's cabin stay.

Smallpox and other imported diseases killed as many as 90 percent of the local native peoples in devastating plagues between 1614 and 1633. Battles with Narragansetts, Micmacs, and Mohawks also took their toll. The founders of Concord outlawed long hair, traditional medicine, and other signs of indigenous culture among their native neighbors. And when illegal land grabs and abuse from settlers triggered a brief rebellion known as King Philip's War in 1675, even the peaceful "praying Indians" of Concord were imprisoned in winter on an island in Boston Harbor, where many died of starvation and exposure. So great was the loss of life in that terrible century that new settlers from Europe marveled at the unoccupied fields and clearings they found ready to be taken over with minimal effort.

Apart from the sprawling urban centers, New England has become greener since Thoreau's century because many of its forests recovered after Americans began to use fossil fuels rather than wood as a primary source of energy and raw materials. Sadly, we have been much harder on the lakes as our numbers have grown and new technologies have emerged, pummeling them with acid rain, nutrient pollution, fish stocking, rotenone poisoning, invasive species, soil erosion, road salt, and climate change. But human impacts on ecosystems are nothing new. The sediment archives of a 6-acre (2.5 ha) kettle lake in Ontario show what might have happened to Walden centuries earlier if the soils around it had been more suitable for farming.

Crawford Lake lies amid fertile, rolling terrain 30 miles (50 km) southwest of Toronto. About 750 years ago, early Iroquoian peoples of the region began to plant crops near their bark-covered dwellings, a new way of living that had originated in Central America hundreds of years before. It transformed their culture, their land, and their lake.

In 2001, researchers from the United States and Canada collected sediments from the bottom of Crawford Lake using a freeze-coring device. Their method involved lowering a metal chamber full of dry ice and alcohol into the lake bed and allowing the extreme cold to flash-freeze a cross-section of mud to its surface. Freeze-coring is designed to capture fine laminations in soft sediments, and the uppermost 2 feet (60 cm) of Crawford Lake mud are full of them. The banded layers, called varves, represent seasonal changes in the lake. The distinctive light bands are deposited in summer when abundant plankton growth changes the chemistry of the water and allows limy mineral crystals to fall out of solution. Below the transition zone burrowing worms, insect nymphs, and the occasional foraging fish had churned and blurred the sediments, but the appearance of undisturbed varves above it signaled the eradication of the bottom-dwellers. The cause? Suffocation. The bottom water lost its dissolved oxygen because of a super-abundance of plankton, a process known as "eutrophication."

The oldest varves in the Crawford Lake core contained the pollen of beans, squash, sunflowers, and maize. It was at first surprising to find so much of it in the mud because the first three crops release their pollen mainly to bees and other insects, and wind-borne maize pollen is too heavy to drift far from its host plants. Then a researcher noticed that most of the pollen lay in strange-looking blobs of organic matter that DNA analysis showed to be the droppings of wild geese. The geese had inadvertently picked up leftover grains of pollen while gleaning seeds from the fields and then plopped them into the lake. The mix of bird dung, human wastes, and topsoil that washed in from the shoreline was like manna from heaven to the microbial communities of Crawford Lake.

Life-giving phosphorus atoms passed from crops to people and geese, and then on to the plankton. The microbes competed among themselves for the bonanza, with cyanobacteria and slender, fast-growing diatom algae such as *Asterionella* and *Synedra* winning out. Blizzards of depleted and dying cells fed bacteria in the darker waters down below, consuming much of the oxygen and smothering bottom-dwellers. The water in the upper layers of the increasingly eutrophic lake became so thick with plankton that many of the bottom-dwelling diatoms in the shallows were also shaded out by the greenish clouds above them.

The Crawford Lake story warns that such human-driven changes can be permanent, an insight that we could only have guessed at from typical short-term ecological studies. The cores show that a eutrophic plankton community lingered for hundreds of years after the settlement was abandoned. Scientists speculate that the loss of dissolved oxygen triggered a self-fertilizing cycle, known as "internal loading," within the lake that released phosphorus from the bottom muds and kept the algae growing. This may also have primed the lake for more eutrophication when Anglo-Canadian settlers began to clear and farm the watershed again during the mid-1800s. Even today, the lower half of the lake still lacks oxygen in summer, a testament to the close connection of prehistoric people to the natural world that was later intensified by the Canadians who replaced them.

We shouldn't be surprised to learn that early indigenous peoples sometimes changed their surroundings in ways that we find undesirable. Like any of us and any other species, they consumed resources and produced wastes, and the intensity of their impacts depended more on their numbers and lifestyles than on some presumed isolation from nature. Most lakes in the Americas contain little or no sign of pre-Columbian people in their sediments, but there are more exceptions to that rule than Crawford Lake alone.

The Mayan civilizations of Guatemala and Honduras developed maize agriculture long before more northerly cultures adopted it, and their increasingly urbanized societies left indelible marks on their lakes as a result. So much exposed topsoil eroded into Central American lakes between 1000 BC and 900 AD that scientists who work in the region have dubbed distinctive sections of their cores "The Maya Clay." Analysis of a core from one Guatemalan lake suggested that a thousand metric tons of soil were lost from every square kilometer of the adjacent countryside over the course of a single year. Eventually, however, Mayan farmers seem to have learned from the mistakes of their predecessors and found more sustainable ways to live in even larger numbers while losing less topsoil.

Non-agricultural peoples sometimes fouled their local lakes, too. One shallow coastal lake in the Canadian Arctic contains unusually high concentrations of phosphorus because Inuit whale hunters began to process and consume their catches nearby eight centuries ago. Diatoms in a core collected by researchers from Queen's University, Ontario, showed that bits of whale tissue left on the ground fertilized lush mats of moss that, in turn, supported a species of tabular *Pinnularia* diatom that grows well on damp mosses. The sediments also showed that nutrients have continued to leak from crumbling bones ever since the whale-processing stopped more than four hundred years ago.

Most indigenous peoples left few traces on their land or lakes until agriculture became more common during the last millennium or so. But there is one widespread, transformative change recorded in lake deposits that many scientists attribute to early hunters about twelve thousand years ago, although the claim remains controversial. It was a great dying on a scale larger than anything seen in the waning stages of previous ice ages, and one that was strangely focused on large mammals.

North America was once home to mammoths and mastodons, cave lions and camels, giant ground sloths, huge bison, enormous *Castoroides* beavers, short-faced bears, and fierce sabertooth cats. Horses had also evolved there over the ages but they, too, vanished and were not seen again until Spanish explorers repatriated them during the sixteenth century. We often think of big hairy mammoths and mastodons as creatures restricted to icy glacial landscapes, but many scientists believe that they might still roam our forests and prairies today if a much smaller and less hairy mammal had not joined them on the continent.

The case for a human cause is strong. We know from their excavated skeletons that mammoths and mastodons lived in a wide range of habitats and climates from Alaska to Florida. Like other large American mammals, their ancestors survived many previous climatic shifts, and there was nothing to prevent them from migrating long distances to reach favored habitats. The de-icing of Canada also opened up vast new territories full of the same kinds of plants that the herbivores' ancestors had grazed on farther south. The most unique environmental change in North America at the end of the last ice age was the arrival of human immigrants from Asia. We also know that there were skilled hunters among them, not only because they left distinctive spear points and kill sites behind but also because lakes preserve additional evidence of it.

Lakes all over North America recorded the demise of the mega-mammals in the form of tiny spores that blew into the water from dung-dwelling fungi. For thousands of years, some of the largest herbivores the continent had ever seen produced correspondingly large piles of feces that *Sporormiella* fungi thrived on. As those herbivores became less and less common around twelve thousand years ago, so did the *Sporormiella* spores in the sediments of local lakes.

Some lakes even preserve the remains of the animals themselves. I will never forget a lecture that paleontologist Daniel Fisher gave at the New York State Museum in Albany some years ago. He began by describing what appeared to be a rather routine excavation of mastodon bones on the bottom of a pond in the Great Lakes region. He and his colleagues had drained off enough of the water to expose a muddy pit that was littered with old bones and tusks. But why would mastodons end up on the bottom of a shallow pond in the first place? Surely the animals could swim as well as modern elephants, and if they fell through thin ice in winter couldn't they simply break their way back to shore?

Fisher soon noticed that the bones lay in distinct heaps, and that each heap contained similar bones with ribs here, femurs there, and so on. Some of the piles were accompanied by odd swirls of sand, and some also had rotten stubs of wood sticking out of them. Slowly, an astonishing picture emerged from those disparate details. This was no accidental burial. The pond was a paleolithic meat locker.

The sand in the swirls matched that along the shoreline and perhaps represented intestines that had been stuffed with sediment to weigh the cache down. The wooden stubs could be the remains of long poles that served as place markers. But is meat even edible after lengthy storage under water? Fisher deposited raw horse meat on the mucky bottom of a small lake in autumn before it froze over and retrieved it in spring. The chunks developed an unappealing yellowish rind during the intervening months, but when it was sliced away it revealed dripping red flesh that was seemingly as fresh as the day it was submerged. Cold, darkness, and low oxygen concentrations in the mud had kept the meat from spoiling.

If Fisher's interpretation of the site is correct, then winter may have been a time of plenty for mastodon hunters who lived in lake country. Perhaps early visitors to Walden Pond used shallow areas of it as a refrigerator in similar fashion, tending groves of slender poles on the ice through the long, harsh winters of paleolithic America.

To find that early indigenous people could sometimes be as unaware of their impacts on ecosystems as modern people helps to humanize them as well as us. It also shows that environmental impacts are not a uniquely modern problem but largely the result of being subject to the timeless laws of nature.

I RETURN TO WALDEN again in December 2016. The day is unseasonably warm and windless, and the reflected images of clouds are sharp and clear as they glide slowly over the smooth surface. During the past year my students and I have been busy analyzing samples from our cores, but rather than sample the lake today I simply want to sit beside it.

In *Walden*, Thoreau wrote of his desire for a fanciful "realometer" to cut through the "mud and slush of opinion, and prejudice, and tradition, and delusion . . . to a hard bottom . . . which we can call reality." I am here to consult my own version of a realometer—this beautiful lake with its ancient sediment archives and the larger perspectives on life that they inspire.

The low water level has exposed a sandbar at the mouth of the cove, inviting me to walk on it. I hunch down, lean over the water's edge, and let my eyes explore it layer by layer. The surface lies as still as the air, and bright sunlight flickers on and off through gaps in the mirrored clouds. Refocusing on the bottom, I scan the smooth pebbles of gneiss and quartzite amid the sand and imagine them tumbling in glacial rivers. Leaning closer, I wait until the water itself comes into focus and just barely make out a tiny speck of a copepod motoring about in search of a meal or a mate.

I wonder how it would feel to sit here with Thoreau, staring together into Walden Pond. Would we see the same things in it? Probably not, but I suspect that we would enjoy sharing our impressions nonetheless. "Time is but the stream I go a-fishing in," he wrote in *Walden*. "I drink at it; but while I drink I see the sandy bottom and detect how shallow it is. Its thin current slides away, but eternity remains." I feel much the same today. I let my imagination sink down to the submerged sediment layer of this present moment, then deeper still.

There are layers upon layers of stories stacked under this lake, and

Sandbar at the mouth of Thoreau's cove. *(photo by Curt Stager)*

any individual increment of mud is just one of many pages in the epic of human existence. It reminds me that all lives are finite and makes me feel less alone in my own encounters with mortality. The long geological history preserved here reveals a deep human connection to the natural world that also comforts me, one that philosophers such as Emerson who considered people to be separate from nature might not have fully appreciated. Untouched wilderness never really existed in North America, at least not since the large mammals vanished, and a Walden without *Homo sapiens* somewhere in the picture might be cleaner but also as artificial as a swimming pool. Envisioning the sediment records beneath the reflections helps me to clarify this truth and my own connection to the world in ways that words alone cannot.

Echo soundings recently obtained by the Salem State University geologist Brad Hubeny suggest that the deposits beneath the eastern basin of Walden Pond are about 20 feet (6 m) thick. Imagine driving a core barrel all the way through those sediments and then leaning the core upright against the side of a two-story house so the top stands level with the eaves. Now imagine climbing a ladder to measure the entire length of that column, not in units of feet and inches but of lifetimes, each one lasting, say, a conservative sixty years.

To get used to those unusual temporal units, consider some familiar time periods in these terms. Two and one-half such life spans separate us from Thoreau, for example, and only four separate us from the American Revolution. Six or seven life spans take us to the arrival of the Pilgrims in Plymouth and eight or nine take us to the first landing of Columbus in Hispaniola. For many people, those few life spans represent the history of America, but the imaginary sediment column puts that misconception into clearer perspective. It represents more than two hundred successive human lives.

Starting near ground level, a thumb when pressed against the first increment of core history might span the lives of the earliest visitors to Walden Pond who arrived after the kettle lake formed, perhaps thirteen thousand years ago. Stone spear points and other artifacts unearthed in the Concord area and pollen records from other New England lakes suggest that they hunted caribou on what was then a mosaic of tundra and spruce thickets much like those in Arctic Canada today. Two more thumb-widths encompass the lifetimes of the children and grandchildren of those early hunters who now join us in spirit on this sandbar.

Continuing in like manner layer by layer, life by life, we approach chest height in a three-thousand-year period of hunting and gathering half again as long as the stretch of time between the present day and the birth of Christ. This represents the long "Paleo-Indian" period of New England's past that began with the retreat of the last ice sheet.

Slightly more than 2 feet (70 cm) higher above the ground we are roughly seventy-five life spans into the story, when warm, dry climates supported a fire-prone mix of savanna grasses, pitch pine, and oak. Some of the oak pollen in this mud may have made a local deer hunter sneeze, spooking a buck he had hoped to take with a stone-tipped dart when the animal came to the lake to drink. As a member of the local "Archaic" culture, he would have dined often on deer, wild turkeys, and acorns in the surrounding forests without ever having heard of the maize, beans, and squash that would not arrive for another seven millennia, longer than the history of the Egyptian pyramids.

Three-quarters of the way up the core, we reach sediments that were deposited when people of the late Archaic to early "Woodland" cultures used some of the earliest clay pots while they camped and cooked beside the lake three thousand years ago. About 2 feet (70 cm) from the top we are sixteen lifetimes away from the present. The thousand-year-old sediments there contain charcoal from the seasonal burning of forest underbrush and maize fields in a place that only became known as "Concord" a few short inches ago on the mud timeline.

Now, while the imagery is still fresh, ask yourself if people are part of the natural order of things at Walden Pond.

What names did the first maize farmers give to this lake, or the Woodland potters before them, or the Archaic deer-hunters before them, or the Paleo-Indian caribou hunters before them? What did they discuss over breakfast on the shore—if they had it—or daydream about while watching reflections from the sandbar at the mouth of "Thoreau's" cove?

We will never know for sure, but other insights from the Walden realometer help to steady me in this particular moment. Changes come and go, whether from ice age cycles or election cycles, but life itself continues. I am a spark among many in that ancient, evolving flame. I glimpse a reassuring measure of eternity in it, too. Microscopic samples of the worlds that Thoreau and many others knew are still here, resting in soft beds of sediment. Now excerpts from my own lifetime are also being preserved along with them in the ongoing story of Walden Pond.

DEEP-DIVING THE RIVER of time with the aid of lake sediment deposits is not only about asking "what happened before?" It is also a reminder to ask "what comes next?" Today is not Walden's final chapter, and the geological archives will continue to accumulate on top of the layers laid down during our own lifetimes for as long as the lake exists. With 102 feet (31 m) of lake left to fill with compacted sediment to today's water level, what now seems like a long history since the last ice age will represent only the opening chapters tens of thousands of years from now.

As you read this, fresh sediments are burying mementos of our century on the floors of lakes all over the Earth, turning the story of the world we know into future memories. If we could glimpse what is now being written into those sediments, what would we find? One need not refer to an imaginary core in this case, because we have some real ones to consider. Here is what we found in our cores from Walden Pond.

A single finger-smudge of Walden mud contains millions of ornate diatoms, the glassy scales of chrysophyte algae, and golden grains of pollen. Scanning a sample of it under a microscope is like beachcombing with microfossils for seashells. The main focus of our study, funded by the National Science Foundation, has been the diatoms and what they reveal about past climates and water quality.

Being light-harvesters just as plants arc, diatoms who live on the bottom rather than floating in the plankton are restricted to the margins of the lake where the water is shallow enough to let sunlight reach them. When lake levels fall, the shallows move closer to the center of the basin, which lets more bottom-dwellers mingle with planktonic species in the mud at the coring site. Shallow-water diatoms in the layers about 2 feet (50–65 cm) from the top of the longest core suggest that decades-long droughts lowered the level of the lake significantly between 1100 and 1300 AD. The sandbar at the mouth of the cove, which is now exposed only sporadically, was likely a perennial feature then. More important for us today, the dry phase occurred during a prolonged warm spell that has been called the "Medieval Warm Period" or "Medieval Climate Anomaly." In contrast, most climate models anticipate wetter, not drier conditions in New England as our carbon emissions warm the planet in the future. This apparent disagreement between lake history and the computer models suggests that the region may be heading into uncharted climatic territory as human-driven warming reheats the world.

The scientists who first cored Walden Pond found the pollen of European ragweed amid the oak and pine pollen in the sections of their cores that dated to the eighteenth and nineteenth centuries, along with flecks of charcoal from forest fires and the migrant shanties that sprang up alongside the railroad. Silt that washed in from the shoreline told of local woodcutting, homesteading, and farming. However, the remains of planktonic diatoms that sank to the floor of the lake during Thoreau's cabin stay were much like those that had done so for more than a thousand years.

As a scientist, I am supposed to treat the mud from Thoreau's time like any other in the core, and I do. But it still feels special when I slip a glass slide with a sample of it under my microscope and swing the 1,000-power lens into place. The last time I peered through the eyepieces at such a sample, I saw a bright disc of light upon which beautiful geometric shapes were strewn like snowflakes. The microscope lens then became a time tunnel through which I could visit the Walden that Thoreau knew.

My favorite diatoms in the mix were also the most abundant. A *Discostella stelligera* shell or "frustule" looks like a tiny glass button with fine striations around the rim and a starburst of ovoid pores in the center, as the name "disc-star" suggests. Each lovely disc was once the lid or base of a silica box that contained a living cell. The ones directly under my gaze floated in the plankton of Walden Pond a little more than a century and a half ago, trapping nineteenth-century sunlight with golden brown pigments and using it to turn air, water, and mineral elements into new versions of itself. Those very diatoms might have swirled in the wake of one of Thoreau's paddle strokes or floated unseen beneath the reflections he described in *Walden*.

The uppermost inches of the core revealed another planktonic community altogether. When I replaced the "Henry sample" under my lens with one from the late 1930s, *Asterionella* and *Synedra* diatoms were more common in the field of view. They were less ornate than the *Discostella*, and although they, too, were made of glass they resembled knob-tipped walking canes and toothpicks. Those diatoms told of nutrient pollution in Walden as they did at Crawford Lake, and their sudden appearance in the core corroborated Deevey's report of cloudy waters near the swimming beach in 1939.

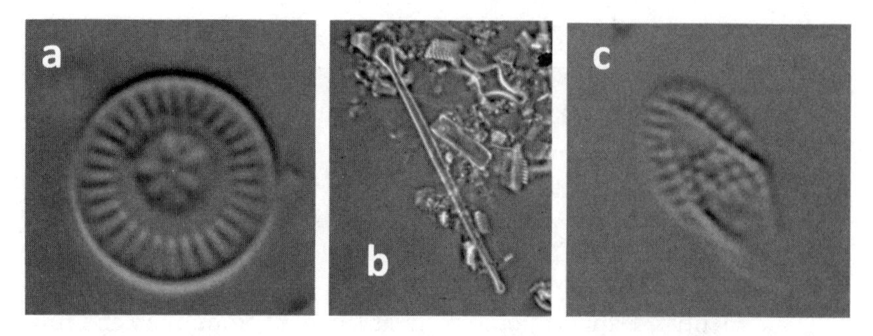

Microfossils from Walden Pond sediment cores. Left: *Discostella stelligera* diatom. Center: *Asterionella ralfsii* diatom. Right: Chrysophyte scale. *(photos by Curt Stager)*

By the 1930s, the rise of automobiles and an increasing stream of visitors had turned Walden into "a community bathtub for greater Boston," as author Barksdale Maynard described it in *Walden Pond: A History*. Nutrients flowed in from swimmer P, possible leakage from the bathhouses and a nearby septic field, droppings from seagulls who fed at an adjacent town dump, and soil inputs from the cutting of a footpath around the lake's easily eroded flanks. In 1957, county officials ordered the clearing of trees and bulldozing of the beach and slopes above it to improve the swimming area, sending nutrient-laden dirt far out into the water and sparking outrage among lovers of the lake. Rotenone treatment and repeated stocking of game fish might also have added enough extra wastes and carcasses to the food web to further enhance the growth of algae.

Swimmer education programs, the closing of the Concord landfill, and efforts to stabilize and revegetate the shoreline have helped to stem the worst of those abuses, and our cores show that *Asterionella* and *Synedra* numbers have more or less leveled off since the 1970s. The *Nitella* meadows soak up much of the human-derived phosphorus that would otherwise support floating algae, but this also means that the lake is potentially more eutrophic than it looks. Low oxygen concentrations in the deep basin in summer warn of a ticking nutrient bomb that just a little more planktonic growth might trigger. Only time will tell if future stewards of Walden treat the *Nitella* meadows as ecological allies rather than mere lure-snagging weeds, or whether a more prolific, human-dependent plankton community becomes as permanent as the one in Crawford Lake.

The remains of chrysophyte algae are also unusually abundant in the sediments of the last few decades. Like diatoms, chrysophytes are golden brown, light-harvesting cells, but they have whip-like flagellae that allow them to swim and thereby remain in the warm, sunlit surface waters of stratified lakes while heavier, less motile algae sink. Chrysophytes have become more common in many lakes worldwide, an increase that some experts attribute to longer summer stratification seasons that accompany global warming. Climate change may help to explain their recent success at Walden, but our influences on the lake have grown so numerous and complex that they have even begun to disrupt one another. The jump in chrysophyte abundances, for example, happened shortly after the poisoning of the fish community in 1968, so it is difficult to tell which factor, climate or food-web effects, was most responsible for the change or what it means for the future of the lake.

The specific nature and timing of these changes are unique to Walden Pond and more extreme than anything else revealed in the last 1,500 years of its sediment records, but they are mirrored in other lake records elsewhere. Today's widespread nutrient pollution, species invasions, extinctions, and soil erosion rival some of the most dramatic environmental disruptions of the geologic past. A growing number of scientists agree that unique signs of our modern connections to nature in the aquatic sediments of the Earth are extensive enough to merit a new "Anthropocene" name for our present epoch. However, they disagree over the date that best represents the transition.

For some researchers, the postglacial extinction of the large mammals makes a reasonable signpost. However, the cause of the die-off is still debated. In addition, it was not truly global in extent because many large mammals persist in Africa, and it occurred over many centuries, so it doesn't make a uniformly clean break in the geologic record.

Others prefer to put the onset at the dawn of agriculture in the Middle East about ten thousand years ago. As farms spread around the world, so too did deforestation, soil erosion, and the environmental effects of urbanism, as Maya Clays and Crawford Lake sediments attest. However, those changes, like the mammal extinction, were not globally synchronous and omnipresent.

In 2015, British geoscientists Simon Lewis and Mark Maslin proposed two candidates for a "golden spike" event to define the start of the Anthropocene. One was a dip in CO_2 concentrations in polar ice cores that dates to the early 1600s when imported diseases killed millions of Native Americans, including many of the Massachusett peoples. So much former cropland returned to forest as a result of the plagues that the sequestration of carbon in new wood and foliage apparently reduced the CO_2 content of the atmosphere.

Their other candidate was the early 1960s, when the world was most heavily contaminated with fallout from the atmospheric testing of thermonuclear weapons during the Cold War. The cesium-137 peak in lake deposits is so widespread that scientists already use it as a time-marker in sediment cores, as we did at Walden. The mud 4 inches (10 cm) below the top of our long core is unusually radioactive, according to a technician at the Museum of Minnesota who analyzed the isotopes for us, a unique signature of the dawn of the atomic age if not the Anthropocene. For the first time in the history of the planet a species has produced an entirely new kind of waste, one that is not simply a rearrangement of preexisting elements but new atoms forged in the violent hearts of artificial stars.

Canadian ecologist Alex Wolfe and colleagues recently summarized the diversity of changes that are revealed in cores from remote back-country lakes worldwide. Artificially generated nitrogen compounds suffuse the recent layers because fossil fuel combustion and the industrial production of fertilizer now dominate the global nitrogen cycle. Arctic lakes that have recently lost their summer ice lids now support planktonic diatoms who are leaving their remains in the most recent sediments for the first time in thousands of years, strong evidence that the recent warming is unusual and not due to natural climate cycles. And a combination of nitrogen pollution and warming seems to be causing a rise of chrysophyte algae in high-altitude lakes from Alberta to the Andes.

Our profound impacts on the natural world occur because we are profoundly connected to it. Unfortunately, those connections are often hidden by the limitations of our senses, as a lake's reflective surface obscures its depths. We can't easily see the teeming atoms of air, water, soil, and organisms that make up our bodies, nor notice what happens to them when they leave us as waste. Nonetheless, these elemental connections hitch us to all life and the Earth itself whether we recognize them or not, just as they did for our predecessors and will continue to do for our descendants. To more fully recognize and respect such connections is one of many challenges we face as the Anthropocene epoch unfolds.

Thin clouds slip quietly over the face of Walden Pond while I prepare to leave the sandbar at the mouth of Thoreau's cove on this December afternoon. I hunch down and brush my fingers through the mirror one more time in a ritual farewell.

We are not separate from nature. We *are* nature, an ancient truth that can perhaps most clearly be seen through the eyes of lakes.

(photo by Curt Stager)

2

WATERS OF LIFE, WATERS OF DEATH

A thing is right when it tends to preserve the integrity, stability, and beauty of the biotic community. It is wrong when it tends otherwise.

—ALDO LEOPOLD, *A Sand County Almanac*

IT WAS A BEAUTIFUL morning in September 1997 when I launched my canoe for a late-season paddle on Black Pond, a secluded lake near my home in the Adirondack Mountains of upstate New York. Maples and birches in the surrounding woods were turning red, orange, and gold, the cool air was fresh and clean, and the sky was a deep, cloudless blue. The colors of forest and sky blended on the surface of the lake amid the noisy splashes of yellow perch who were rising all around me. I had often seen fish feed on emerging insects here, but these fish were not feeding. They were dying.

The Adirondack State Park encompasses 6 million acres and is the largest park in the continental United States, larger than Yellowstone, Yosemite, Glacier, and the Grand Canyon combined. It sits astride a broad, forested dome of ancient anorthosite rock that is famous for the beauty of its rugged peaks and abundant lakes, but also for the ecological

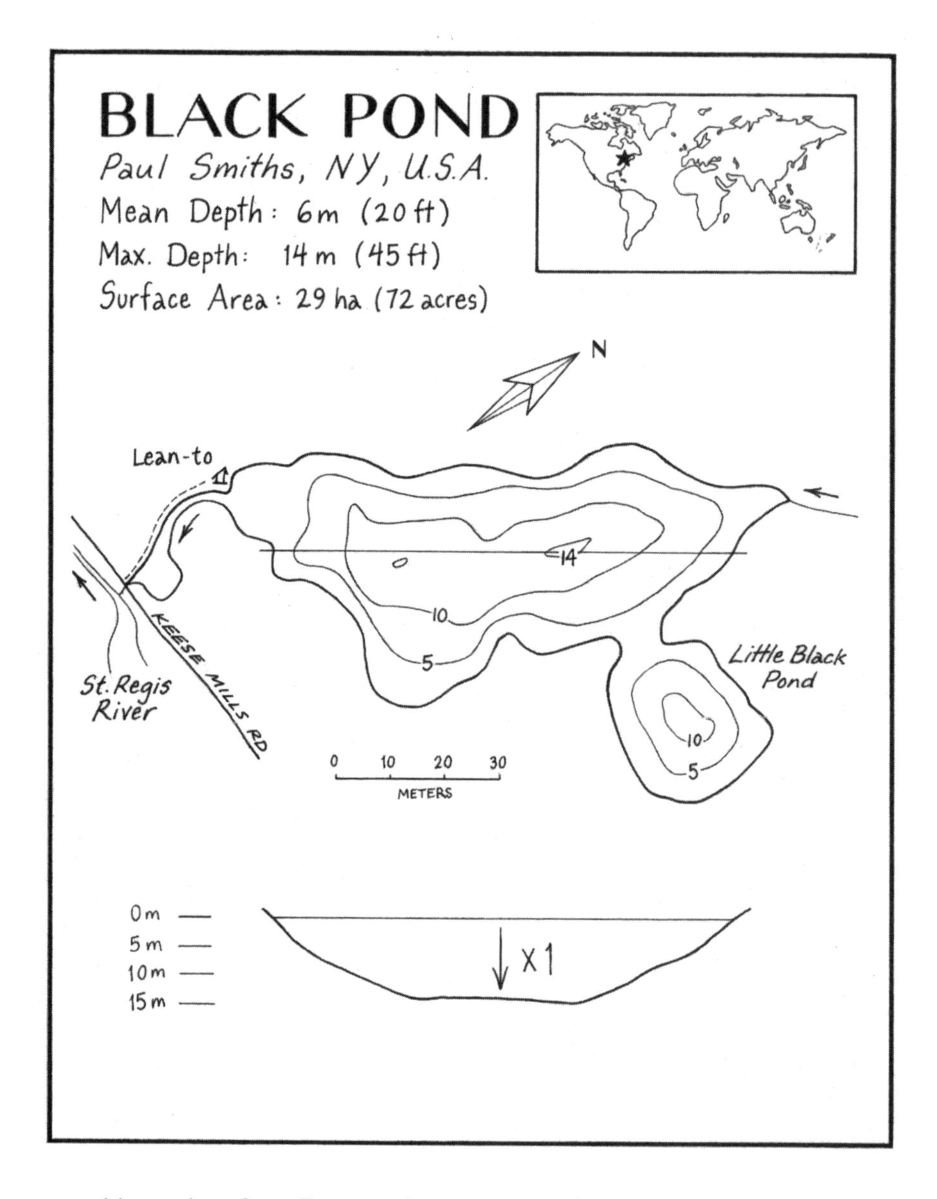

problems they face. During the 1980s, the Adirondacks were Ground Zero for the harmful effects of acid rain, and studies carried out here helped to identify its causes and consequences. Climate change took over national headlines before the struggle against acid rain could be presented as the inspiring and informative success story it was, and when other scientists learn that I study Adirondack lakes they still often ask, "How is the acid rain doing?"

The dying perch in Black Pond, however, were not victims of distant smokestacks and tailpipes. They were being poisoned by the agency that is most responsible for protecting Adirondack waters and their inhabitants, the New York State Department of Environmental Conservation, or DEC.

The story of Black Pond reflects a clash of worldviews that accompanies our efforts to care for lakes and the planet as a whole. It is not a clear-cut saga of good versus evil. All parties involved care a great deal about lakes and see themselves as the "good guys." Therein lie the drama and the tragedy. The poisoning of lakes with fish-killing pesticides such as rotenone in order to purge them of unwanted species is a necessary evil to some and just plain evil to others. I have watched this conflict evolve since I moved to the Adirondacks in 1987, and at times I have been caught in the crossfire. In recounting some of those experiences here, I will also examine a problem that I believe is more fundamental to our future than arguments over whether or how to kill perch. It is the problem of understanding our place in nature in the Age of Humans.

WHERE THERE IS LIFE there is death, and the interweaving of the two produces the web of existence we call ecology. People have always known this to some extent, but as our societies become more urban and compartmentalized we more easily lose sight of our connections to that global fabric. Modern science and technology have played a role in producing the illusion of separation, but they have also uncovered connections that our forebears could only imagine, including the shared components of all life, the atoms.

For many of us, awareness of atoms is still too new and strange to be fully incorporated into our sense of who we are, and if at times we seem to be disconnected from nature it is largely because the individual elements that link us to it are too small to be seen easily. When we know more about them, however, we can learn to perceive them more clearly in daily life through atomic eyes.

I find many such connections at Black Pond on a warm summer day in 2016. To do so, I use a three-step questioning process that can be applied to any ecosystem.

Who lives here?

Where did their elements come from?

Where will their elements go?

It is a short drive along a winding rural road from my office at Paul Smith's College to the unpaved parking lot where Black Pond empties into the outlet of Lower Saint Regis Lake. A barrier dam prevents the larger lake's fish from swimming into the pond and undoing the work of the most recent rotenone treatment, which took place in 1997. A sign beside the parking lot tells visitors that this is a refuge for a unique heritage strain of Adirondack brook trout that the DEC staff collect eggs and milt (trout sperm) from in order to raise and stock their fingerlings elsewhere. In fine print, the sign asks anglers to limit their catch to two or three trout. Nobody is checking the fishing methods today, though. Drought has dropped the water too low for easy boat access through the shallow outlet, even for the nonmotorized watercraft that are the only boats allowed here. I take the quarter-mile footpath through the trees along the shore instead.

The outlet is partially blocked by the dam, so it has pooled up and become a slender pond in its own right rather than a stream. Its margins are thick with white water lilies and the purple, honey-scented blossoms of pickerelweed, all busy with insects. To my left is a steep, wooded ridge shaped by the last ice sheet. On the opposite side of the outlet 50 feet (15 m) away, an outcrop of gray anorthosite, smoothed by the same ice, meets the water and its own reflection. Whirligig beetles and water striders ripple its mirrored image like raindrops. The shallow water close to shore is clear enough to reveal mud-colored minnows cruising above the soft bottom, betrayed only by their shadows. Nothing in this scene would indicate that the lake has been rotenoned five times since the 1950s.

Minutes later, I arrive at a log lean-to shelter in a hemlock grove beside Black Pond proper, a half-mile oval that is encircled by mixed hardwood and conifer forest in various shades of green. I step to the water's edge and begin to list the life forms I can see and hear.

Black Pond. *(photo by Curt Stager)*

Dragonflies hunt among the pickerelweeds. Fat tadpoles rest on the bottom while their parent frogs cluck and grunt from the cover of sedges along the shore, and the silvery voice of a hermit thrush echoes in the woods behind me. Farther offshore, a loon dives down to where cold-loving trout prefer to lie at this time of year, leaving the shallows to the minnows. All of these animals are seeking food, so there must be smaller organisms to see as well.

I lean closer to the water and extend my hand to steady myself on a rock among the sedges. A slippery green frog leaps out from under it with a yelp, and my heart leaps, too. I reach into the cool depths and turn over a flake of bark on the bottom. Many-legged isopods and a fork-tailed mayfly nymph scuttle for cover. When I bring my face to within a few inches of the glassy surface, baby water striders no larger than poppy seeds skitter deeper into the sheltering vegetation. All are potential lunch for predators.

Peering through the surface film I see no obvious signs of smaller creatures. I know that they live here, though, because I have caught them in plankton nets. I envision zooplankton gathering in diffuse swarms farther out in the pond and deeper down where it is more difficult for their predators to see them during daylight hours.

That list will do for starters. I select a minnow for the next question: Where did the elements within this little fish come from?

The minnow appears to be a blacknose dace, and I presume that he or she often dines on aquatic nymphs and forest insects who fall into the water. They disintegrate into molecules inside the dace, enter the bloodstream, and break down further within countless cells. The atoms within those molecules are then rearranged into enzymes, hormones, body fat, fins, scales, or anything else the dace needs. The diverse insects of the water and air thereby become this minnow, not merely in some poetic sense but literally. Blacknose dace are recycled bugs.

The conversion of bug flesh into fish flesh is possible only because bugs and fish are made of the same interchangeable elements. It is as though they were built from Lego bricks that can be recombined into various shapes over and over again. The same few kinds of elemental Legos can also build every other animal in and around Black Pond, including human beings. It is this mind-boggling fact that most decisively lays to rest any idea of us being separate from nature. When we are aware of the atoms that comprise and surround us, such a claim seems as silly as a comic book snowman who denies his elemental connection to water.

Many of the atoms of the dace's latest meal originally entered the food web of Black Pond through diatoms and other algae. I try with difficulty to think of the minnow as processed pond scum as well as recycled bugs. And before the algae? Therein lies the most profound transformation of all, the almost miraculous conversion of inanimate matter into life on Earth.

Soft green mists of algae surround a submerged twig, and when I lean closer over the water's edge I see slim filaments glistening like cobwebs. The sunlight that warms the back of my neck is also helping the slender threads to weave dissolved minerals and gases into their cells. In the creation story of the Old Testament, the first human was made from wet sediment into which the breath of life was blown. A similar process is happening here and now among these algae. When I focus instead on the reflection of my own face in the still surface of the lake, I see the elements of air, water, and earth there, too.

In Black Pond as in any waterscape or landscape on Earth, every species is more or less connected to every other by shared atoms. It is that kind of interconnectedness that scholars invoke when they apply the prefix "eco-" (from the Greek *oikos*, dwelling or family) to terms like ecology, ecosystem, and economy. The jargon that scientists use to describe the flow of matter and energy between living beings and their surroundings is also full of references to *trophos*, the Greek term for "one who nourishes or feeds." Algae and plants build their bodies from inanimate scratch and are therefore called autotrophs, or "self-nourishers." Animals are heterotrophs (nourished by others) who must obtain their elements and energy from other living beings instead. Eating is body-snatching, and in our case it also involves the classification of body parts as food.

Some languages reflect the *oikos* and *trophos* of our meals more explicitly than others. Swahili, the historic trade language of East Africa, uses the term *nyama* to refer to wild animals in general and also to meat. In contrast, English speakers more often use their food-language as a shield against what they are actually doing. For many of us, chicken is not a bird or its muscle tissue but a main course or packaged product from a supermarket, an inanimate substance whose name is as meaningless as if it were spelled as it is pronounced, "chkn." A hog's muscles become "pork" in a package, and "dairy" is white stuff in the fridge rather than squeezings from the mammaries of confined cows.

Those tongue-in-cheek descriptions would of course be unattractive on menus or grocery labels, but they illustrate an important point. Verbally obscuring our elemental connections to other living things makes it more difficult to recognize them. To live without understanding the ecological significance of eating and drinking is not only to be out of touch with nature but also out of touch with reality.

Although the diversity of life forms at Black Pond can seem like a jumble of moving parts, there is a basic structure to it that applies to ecosystems and economies alike. For the system to persist, someone has to provide the raw materials to operate and sustain it. Algae, cyanobacteria, and plants represent those crucial roots of the web that animals depend upon for resources that they cannot harvest directly from air, water, minerals, or sunlight. Where autotrophs thrive in great abundance, the term *trophos* pertains in a more general way. Limnologists refer to a lake that is cloudy with algae or thick with plants as eutrophic ("well nourished") as opposed to a less productive and cleaner-looking oligotrophic lake or a middling mesotrophic lake.

The living community of Black Pond emerges from algae on the bottom and in the plankton, as well as from pickerelweeds in the shallows and even from the surrounding forest. Lakes are closely tied to their surroundings, and Black Pond's submerged trophic web is nourished by soils and bedrock, leaf litter, rotting logs, airborne pollen, and the atoms of land insects that birds, frogs, and fish may release into the water after consuming them. The flow of matter and energy also runs

Black Pond outlet reflections. *(photo by Curt Stager)*

back to the forest. Between a quarter and half of the energy needs of certain woodland birds are supplied by insects who cross the air-water interface, and a hermit thrush may carry the atoms of emergent aquatic insects in his or her feathers before death and decay release them among the hungry roots of trees and wildflowers.

As for the question of where my chosen dace's atoms will go next, a trout or loon is a likely destination. When rising water levels later bring anglers back to Black Pond, some former fish atoms might become parts of people. The air that the anglers inhale during their stay will also contain oxygen produced by algae in the water beneath them, part of the same oxygen supply that sustains the fish they pursue. It will become metabolic water inside their cells and reemerge in their breath's vapors, perhaps to soak back down into the lake again. The atoms of all organisms eventually become part of something else until digestion or decomposition scatters them where autotrophic gatekeepers might funnel them back into the realm of the living yet again. Individuals come and go, but our more widely dispersed atomic selves are virtually immortal.

The fabric of life is bound by atoms and light to the planet and its nearest star, an age-old truth that inspired naturalist John Muir to write, "When we try to pick out anything by itself, we find it hitched to everything else in the universe." In lakes of the Adirondacks as in most others around the world, the trophic web is now being picked at from many directions by one of the most influential members of the global ecosystem, *Homo sapiens.*

THE WATER IN Black Pond is normally brown with tannins and other carbon compounds that wash in from the surrounding woods and soils, and on misty mornings it resembles a steaming bowl of forest tea. During the early 1990s, however, the water was pea green with cyanobacteria. Like plants and algae, cyanobacteria are solar-powered autotrophs who often produce useful substances that are in great demand in the food webs of lakes. In the case of *Anabaena*, however, too much of a good thing can mean trouble for fish and other animals.

When cyanobacteria become superabundant in a lake, other bacteria in the rain of decaying cells beneath the murky surface may consume so much oxygen that they force fish to surface or suffocate. *Anabaena* can also produce unpleasant odors and tastes because they protect themselves from grazers among the zooplankton with toxins that can sicken or kill. If enough *Anabaena* thrives in your lake, you might not be able to use the water for drinking or even for swimming.

Anabaena blooms can also exhibit a form of group behavior that I first noticed while canoeing on Lower Saint Regis Lake. On that day I had enough time between lunch and my afternoon teaching duties to paddle from the campus boat launch to the far shore and back. The water was muddy green when I crossed the lake, but before I was halfway back the bloom had vanished. The sudden change felt as eerie to me as if the microbes had conferred among themselves and decided to hide. My colleague, Corey Laxson, later explained what had happened.

Under a microscope, *Anabaena* resembles a strand of brownish-green beads. Each bead is a light-harvesting cell equipped with gas bubbles, or "vesicles," that help the strand remain afloat. The vesicles give *Anabaena* a competitive advantage in warm, stratified surface waters, particularly over diatoms who are weighted down by their glassy shells and must be stirred back up into the sunlight by currents. On bright sunny days, however, the photosynthesizing cells can become so overpacked with newly made sugars that the bubbles burst and the strands sink until new vesicles form. According to Corey, the disappearance of the bloom was caused not by a conscious migration but by a mass deflation of internal balloons.

To understand why Black Pond became a cyanobacterial paradise it is necessary to identify sources of key elements that could favor *Anabaena* over other phytoplankton. Only five kinds of atom top the list of elements in living things. The big three are carbon, hydrogen, and oxygen. They abound in air and water and are therefore fairly easy to come by. The other two, phosphorus and nitrogen, are the usual suspects when a lake goes green because they are more difficult to obtain.

Animals and other heterotrophs take their P and N from other organisms, but most autotrophs have to scrounge them more creatively from raw materials. Phosphorus erodes very slowly from certain kinds of

Phytoplankton blooming in a eutrophic Adirondack lake.
(photo by Curt Stager)

rock and sediment and is the rarest of the main elements in ecosystems. Algae and cyanobacteria snatch it from solution as soon as it arrives in runoff, groundwater, wastes, or decaying cells. Nitrogen is the dominant gaseous element in lake water as well as the atmosphere, but only a few species of bacteria can "fix" it into useful forms. *Anabaena* filaments do it in enlarged, bead-like heterocyst cells, which means that they can extract four of the five most critical elements directly from the water around them. Adding the final growth-limiting ingredient of phosphorus to a lake can therefore hand its phytoplankton community over to nitrogen-fixing cyanobacteria.

One of the first scientists to recognize the role of limiting nutrients in ecosystems was the German chemist Justus von Liebig. During the mid-nineteenth century he proposed that plant growth in gardens and farm fields is limited not by the total amount of resources available but by the rarest resource. His Law of the Minimum was later popularized through the concept of "Liebig's barrel," in which an ecosystem is likened to a wooden tub that is built from staves of unequal length. When water is added to the barrel it can rise only to the level of the lowest stave in the rim and no further. In many lakes, phosphorus is that limiting stave.

Where did Black Pond's extra phosphorus come from? Common sources include soil erosion; fertilizer from lawns, farms, and golf courses; leakage from septic systems; or, as in the case of Walden Pond, leakage from swimmers. Upper Saranac Lake, located several miles from Black Pond, developed cyanobacterial blooms in large part because a DEC fish hatchery released wastes into a tributary of the lake before the error was corrected. No such P-sources exist at Black Pond, however.

A lake's own sediments can be another potential source of P atoms. When the bottom waters are well oxygenated they tend to rust iron-bearing minerals in the mud and form a layer that traps the elements of old dead plankton beneath it. If the oxygen supply runs low, as can happen when a lake stratifies during a long, hot summer, the rusty cap dissolves and allows buried phosphorus to escape into the water. If climate alone were the cause, however, other lakes in the area should have bloomed along with Black Pond, and they did not.

Yet another possibility emerged when I learned that Black Pond had been repeatedly treated with rotenone. Fish are so full of phosphorus that Native American farmers and Anglo-European settlers traditionally used them for crop fertilizer. If rotting fish can fertilize soil, then couldn't they turn a lake eutrophic as well? Later, the bottom sediments might also become a long-term source of recycled phosphorus as they did at Crawford Lake.

With this last possibility in mind I began to investigate rotenone, too. It is derived from a South American plant extract that indigenous people have used for centuries by dropping it into streams and collecting fish who float to the surface. The poison enters the bloodstream of a fish through the gills, but humans and other creatures who lack gills are apparently unaffected. Rotenone breaks down in a lake within days to weeks, and the DEC usually applies it in autumn because the cooler temperatures give it more time to do its work before it decomposes. Limiting its use to autumn also avoids the amphibian breeding period in spring, leaves plenty of fish in the lake for loons to raise their chicks on in summer, and avoids the main tourist season when hikers, boaters, and anglers are most likely to notice. By the time winter is over the rotenone is no longer a threat to new fish who are stocked in the lake.

People who know about the practice tend to love it or hate it, and there are solid arguments and strong emotions on both sides of the debate. Proponents note that nonnative species can be a threat to native fish and that rotenone treatment is an efficient, cost-effective way of removing them. The DEC calls the practice "reclamation" in order to highlight the goal of taking a lake back from alien invaders, and more than one hundred Adirondack lakes have been reclaimed since the 1950s.

The DEC originally treated lakes with rotenone in order to support the stocking of popular nonnative fish such as rainbow and brown trout, but more recently it began to protect native varieties of brook trout in that manner, as well. Nine genetically unique heritage strains have been identified in the Adirondacks thus far, and most of them are threatened by invasive competitors or predators such as golden shiners and northern

An Adirondack heritage brook trout from Windfall Pond.
(photo by Curt Stager)

pike. Water quality problems, climate change, and overfishing also take their toll. Without reclamation to help fisheries managers support them, the reasoning goes, heritage trout might not survive.

Opposing the practice are environmental groups who object to the mass killing on moral grounds or worry that rotenone might have unintended effects on the ecosystem. For them, reclaiming a wild lake is like clear-cutting a forest while hiding the damage under water. Few if any lakes are studied in sufficient detail to document the full spectrum of ecological impacts when they are reclaimed, and the traditional focus on commercially valuable game fish suggests that other species might be unintentionally harmed or lost.

Such concerns are not unfounded. When Dragon Lake, British Columbia, was "rehabilitated" (the Canadian version of reclamation) half a century ago, fisheries staff found two species of whitefish among the carcasses. Both were new to science and unique to the lake, perhaps the first fish species ever to be identified by being driven to extinction. Similarly, the officials who had Walden Pond poisoned in 1968 to boost sport-fishing ignored Thoreau's reports of distinctive forms of bream and other species who might have represented genetic strains unique to Walden.

I became more deeply involved in this issue in 1997 when the DEC proposed to reclaim Black Pond again. They wanted to remove yellow perch who had once again dominated the fish community and to replace them with heritage brook trout. Perch are classified as an invasive species in the Adirondack uplands, largely on the basis of a nineteenth-century survey that found none in the high country. The hardy, prolific fish have since become widespread in the region through formal and informal stocking and immigration by way of interconnecting streams. Perch compete with brook trout for food and eat their eggs, and they can exclude trout from a lake under certain conditions. Nonetheless, the recent *Anabaena* blooms in Black Pond also made me wonder if repeated reclamations had overfertilized the sediments with phosphorus. If rotting fish can make a lake susceptible to cyanobacterial blooms, then long-term effects of reclamations on water quality might threaten heritage trout rather than help them.

The DEC was required to obtain permission from another branch of government, the Adirondack Park Agency (APA), before reclaiming lakes in the park. The APA usually approved such requests, but this time public pressure made them hesitate. Resistance from opponents during a recent reclamation effort had turned ugly and made headlines. A DEC spokesperson called it an assault on fisheries officers, an attorney for a protester reported that his client was bound to a tree in a hugging position and made to slide down it while his face scraped against the trunk, and an environmental group threatened to sue the APA for improperly issuing the permit. Before ruling on the DEC's latest request, the APA announced a public hearing in May and invited Paul Smith's College president Pete Linkins and me to offer a scientific perspective.

At the hearing I expressed my concerns about possible water quality effects and the shortage of detailed, long-term studies on the ecology of reclaimed lakes. An APA official proposed that the reclamation be allowed to proceed as a learning experience in which students and faculty could work with DEC staff to study the lake before and after the rotenone treatment. He then asked my boss if a faculty volunteer would help to organize and execute such a study. The president shot me a meaningful glance that overruled my previous summer plans.

After several meetings between the college and the DEC, the project was ready to roll. My team would monitor and document the ecosystem throughout the coming year. The next autumn, the reclamation would be carried out and then followed by several years of intensive studies to document any changes or lack thereof. Students would gain valuable experience, the trout would get their breeding pond, and sound science would test assertions on all sides of the issue. It seemed like a win-win situation until September, when I received a phone call from a colleague. The DEC was about to reclaim the pond before our survey could begin.

Looking back on the situation now, I can't recall what signs the fisheries staff might have given that they would bypass the initial study once their permit was issued, or why nobody launched a legal challenge. At the time it seemed like the only remaining option was to witness the reclamation and learn from it what we could.

To be clear, I am not necessarily opposed to killing fish with rotenone, especially in artificial reservoirs that were created for recreational use. Even the wildest-looking Adirondack lakes are not completely pristine either, having been exposed to air pollution, invasive species, and other human impacts, and if unique native trout need a refuge then it makes sense to me to clear a place for them when necessary. In addition, I understand the importance of providing sport-fishing opportunities that lure people outdoors and bring much-needed dollars to the North Country. However, I also believe that we should fully understand what we are doing if we decide to "save" a lake with rotenone and ensure that the decision stands up to scientific and ethical scrutiny.

The saga of Black Pond reflects the complexity of our interactions with lakes and we will return to it shortly, but another Adirondack water body is worth considering first in order to gain a wider perspective on what might constitute a desirable lake and how best to care for it.

WHEN I FIRST visited the Adirondacks during the 1980s, Bear Pond was a secret beauty spot that locals dazzled outsiders with. To reach it a canoeist had to paddle across Upper Saint Regis Lake, carry uphill, cross a smaller and boggier pond, and then carry again over a narrow divide to the forest-rimmed shore. I never had much luck fishing in it but that's not what most of us went to Bear Pond for in those days. Its water was spectacularly blue and transparent as a swimming pool. A large, smooth boulder that was shaped like a loaf of bread made a prime diving and sunning platform offshore, and many a date or visiting friend was successfully impressed by a visit to lovely Bear Pond. Most of us did not realize, however, why its water was so invitingly clear.

Pure water registers a neutral 7 on the pH scale of acidity. The pH of Bear Pond was closer to 5, roughly one hundred times more acidic. The lower the pH of an acidic lake the fewer species can survive in it, and Bear Pond's clarity meant that acidification had left it nearly devoid of plankton, leaving little or no food web in the open water. Sulfuric and nitric acids in the rain had also leached aluminum ions from forest soils, so any fish who managed to survive the shortage of food in such a lake also had to respire with gills that were weakened by metal toxicity. For trout it could mean a slow death by starvation and suffocation.

The contaminated air was harming more than fish alone. Adirondack spruce forests were beginning to sicken and die, and people were also being affected. I remember when gray haze often blocked views of the high peaks around the village of Lake Placid during the 1980s. I thought it was merely due to summer humidity until a knowledgeable friend corrected me. "It's pollution from the Midwest," she explained. "When sunlight hits the fumes it turns them into toxic smog." On that particular summer day, lung-damaging ozone concentrations in the High Peaks Wilderness Area were higher than in Los Angeles, and hikers were being cautioned to beware of the health risk.

The fierce public debate that arose over acid rain had much in common with today's confrontations over global warming. Scientists identified coal-burning power plants and motor vehicle exhaust as major sources of the problem. When concerned citizens and environmental organizations pressed for stricter controls on those fossil fuel emissions, industry executives and their political allies resisted. Many of them accused their opponents of trying to destroy the American economy and denied that air pollution was acidifying lakes at all.

Into that debate came a group of experts armed with new techniques that helped to demolish the ill-founded arguments against pollution controls. Scientists had long used sediment cores to study lake histories, but they usually ignored the most recent mud layers because they were too soft to collect without disturbing them and difficult to date. Radiocarbon ages on sediments younger than three hundred years or so are unreliable, but new dating methods using lead-210 and cesium-137 allowed recent sediment chronologies to be reconstructed in detail. Novel sampling methods captured the loose top layers intact, and powerful statistical tools allowed pH changes to be reconstructed from diatoms and other algae in cores. The much-improved studies showed that many Adirondack lakes had indeed acidified in response to fossil fuel pollution, and Bear Pond was one of the lakes that told the tale.

This is the point in the story at which most of us have lost track of the narrative. During the 1990s global warming distracted the mainstream media, and acid rain was largely forgotten by the general public. So sudden and complete was the attention switch that monitoring equipment for research aimed at acid rain in the Adirondacks was abandoned in the woods for lack of grant funds. In turning away from that story, many of us missed its most important chapter.

Supported by the sediment core studies and other research, Congress updated the Clean Air Act in 1990 to reduce acid emissions. The United States Environmental Protection Agency (EPA) worked with state governments and electric power companies to establish a cap-and-trade system and encourage cooperation among regions that produced the majority of the pollution and those that suffered the most from it. Pollution-scrubbers were added to power plants and vehicles without seriously harming corporate profits. As a result, sulfur and nitrogen oxide emissions declined significantly in the years that followed, smog cleared over cities and wild lands alike, and many formerly acidic lakes began to recover.

The success story of acid rain demonstrates that good science combined with strong, thoughtfully crafted legislation can solve major environmental problems without ruining the economy. In the case of acid rain, federal oversight was required to guide the disparate parties toward a reasonable and effective solution. Now that we face the larger problem of heat-trapping carbon emissions from the same fossil fuel sources, our previous success in dealing with acid rain shows how we might succeed again with similar approaches despite similarly politicized denials and delays.

That was the acid rain story in a nutshell, but a new twist in the tale recently emerged at Bear Pond. It is now no longer the alluring destination it once was, not because efforts to save it from acid rain failed but because they succeeded. The lake is less than half as clear as it was during the 1980s and more brown than blue, partly because warmer, wetter climates are now leaching more organic matter from local soils and partly because the pH of the water has become more suitable for phytoplankton. Ironically, Bear Pond's recovery has therefore made it less attractive to visitors.

Herein lies some of the difficulty in becoming responsible stewards of ecosystems. When a lake changes in ways that we dislike we naturally want to fix it, but environmental restoration is often easier said than done. Our sense of what is best for a lake is shaped by perceptions that may or may not accurately reflect reality, and the language we use to explain our actions often reveals more about ourselves than about lakes.

At the height of the acid rain controversy, for example, environmental activists often described clear acidified lakes as "dead." At the same time, algae-rich eutrophic lakes were also being called "dead" amid efforts to curb phosphorus pollution. The terminology was well intentioned but misleading. By definition, eutrophic lakes are full of life even though fish may suffocate in them. Oligotrophic acid lakes may be fishless but they can nonetheless support acid-tolerant algae, plants, and small creatures on the bottom where sunlight penetrates the clear water. In other words, eutrophic lakes are full of floating life and oligotrophic lakes can be full of bottom-hugging life. In both cases, the "dead" labels simply reflected a focus on commercially valuable fish because most of us care about them more than about plankton or insects. We may therefore imagine that we are restoring a lake for its own sake when in fact we are doing it most of all for ourselves.

When lake managers "reclaim" or "rehabilitate" a lake with rotenone, their choice of terms reflects their worldviews in similar fashion. For example, one study conducted in the Adirondacks during the 1990s concluded that reclamation restores the biotic integrity of lakes. The study was flawed for several reasons, including its short duration and inclusion of a previously reclaimed lake among supposedly undisturbed control waters, but it also demonstrated the use of language to convey an image of lakes as servants of humanity. If we value lakes mainly for angling, then replacing undesirable fish with more popular species can seem to restore the biotic integrity of a lake when, in a broader ecological sense, it does nothing of the kind.

Angling and fisheries management represent important and time-honored human connections to the natural world for millions of people around the world, but there is more to lakes than sport fish. My wife,

Kary, recently suggested a useful way of describing opposing views on lake management based on her training as a musician. "It's like playing music on different instruments," she said.

In the case of Black Pond, focusing fisheries management efforts on brook trout to the exclusion of the rest of the food web is like playing a melody on only the white keys of a piano. Timing a reclamation to reduce risk to charismatic amphibians and loons adds a few more intermediate notes on black keys, and the project is clearly defined in black and white, as it were.

In contrast, many limnologists prefer a broader approach in which all members of an ecosystem are potentially important. Rather than limiting the musical score to the few notes available on a keyboard, they also recognize additional notes between the keys. In this case, imagine using a more adaptable violin to play music that is richer in microtones like that of the Middle East.

Now imagine presenting the two styles together for a mixed audience and envision the response. To the limnologists the keyboard music is fine but too limited to be satisfying. To the fisheries managers the notes played between the keys might sound strange and unpleasant. After the 1997 reclamation at Black Pond, my follow-up studies triggered a range of responses, much like the hypothetical clash of musicians.

ON THE DAY of the reclamation, the lean-to on the shore of Black Pond became a busy staging area. The plan was to use long hoses to poison the bottom waters first and prevent perch from seeking refuge there, then purge the surface layer of survivors. When I arrived, gray plastic buckets of Noxfish, a commercial brand of rotenone, were being emptied into a barrel that was mounted crosswise in a small motor launch. Beside them stood tins of acetone, a carcinogenic solvent that was mixed with the Noxfish to help it dissolve and disperse. Two DEC employees then pushed off from shore and began to crisscross the lake in their rotenone-dispensing boat, trailing creamy chemical stripes behind them.

A friend and I cruised the shore in my canoe, staying clear of the DEC craft. Fish were already splashing in the middle of the lake, but

Reclamation of Black Pond, 1997. *(photo by Curt Stager)*

the surface revealed only part of what was going on down below. Peering into the water near the eastern shore, we watched dark masses rush silently back and forth beneath us like the shadows of deranged clouds. They were schools of desperate perch trapped in a lake that had become an extermination chamber. By late afternoon the shallows were choked with dead fish.

The following week, the water was cloudy and rank with decay. Several weeks later most of the murk and odor were gone but my plankton net still captured no copepods or cladocerans, who normally thrive during the autumn mixing period. Instead, phantom midge nymphs prowled the open water. The worm-like predators on zooplankton can burrow into bottom sediments when necessary and probably escaped most of the poison that way. They rarely swim near the surface during daylight hours because fish see and eat them, but for the time being Black Pond was all theirs. From their perspective the lake had been reclaimed from the tyranny of fish, but their new setting was also low on provisions.

By the following summer the newly stocked heritage trout were settling in nicely. Copepods and cladocerans were abundant again, no midge nymphs were to be seen, and there was relatively little *Anabaena* in the plankton. Without a complete pre-reclamation survey from the early twentieth century for reference, there was no way to know how fully Black Pond's food web had recovered, but the diatom community was easier to evaluate. My students and I collected several sediment cores that allowed us to compare today's resident diatoms to those of the past. What we found raised new questions about the state of the lake.

Before the first reclamation in 1957, the planktonic diatoms in Black Pond were typical of mesotrophic Adirondack lakes, mostly heavy, round *Cyclotella*. Soon afterward buoyant, needle-shaped *Asterionella* and *Fragilaria* dominated the samples amid decay-resistant heterocysts of *Anabaena*. The plankton community had suddenly shifted to a more eutrophic condition and stayed that way through repeated reclamations, stockings, and perch colonizations. In this case, the reclamations did not restore the biotic integrity of Black Pond because the phytoplankton community did not return to its previous state. The fish community also remained unrestored because native lake trout and the original resident strain of brook trout had been eliminated by the first reclamation.

Our cores demonstrated that the eutrophication of Black Pond occurred during the 1950s, but they also sparked disagreements over its cause. The dispute centered on two related but opposing mechanisms for changing a lake through "trophic cascades," ecological terminology for influencing a food web from the bottom or the top. Trophic cascade theory holds that feeding nutritious elements to algae or plants at the base of a food web can affect consumers higher up, but also that changes among consumers can affect the producers down at the base.

The eutrophication closely followed the first rotenone treatment, so it could presumably have been a bottom-up effect. Phosphorus from dead fish might trigger cyanobacterial blooms, and the resultant loss of oxygen could unleash even more P from the sediments. The stocking of several thousand hatchery fish each year, many of whom would die and decay in the lake, also added extra phosphorus to the food web between reclamations.

Alternatively, the shift might represent a top-down trophic effect of yellow perch colonizations, a view favored by the DEC. Young perch eat

Aftermath of reclamation at Black Pond, 1997. *(photo by Curt Stager)*

zooplankton, who feed on phytoplankton, and by consuming those tiny grazers the perch might have left more of the algal crop unharvested, turning the lake green. The DEC staff maintained that the eutrophication was sustained by perch recolonizations that occurred after each of the reclamations.

Because the first perch arrival and rotenone treatment both reportedly happened during the 1950s shortly before the eutrophication, there was no way to tell from sediment cores which explanation was correct—or so we thought.

WELL BEFORE the results of our core studies were published, I had already become unpopular among the DEC staff. Through friends I learned with regret that my questions and investigations were seen as attacks on their life's work and their profession as whole. I knew and liked the staff, having befriended several of them in the past (which is why I do not give their names in this account), but we worked in different cultures. They were doing their jobs as traditional DEC fisheries managers, but my job as a scientist and educator was to question assumptions and to study lakes not only as economic resources but also as ecosystems.

Relations worsened further when new findings challenged two additional assumptions, that Adirondack reclamations have posed no risk to humans and that perch are alien invaders.

After the last reclamation, I learned that Black Pond had been treated with a more persistent poison, toxaphene, in 1963. Unlike rotenone, toxaphene is a synthetic cocktail of more than six hundred chemicals that affects all kinds of animals, not only those with gills. It also persists in lakes much longer than rotenone. Its use as an agricultural pesticide was discontinued in the United States during the 1980s due to human health concerns, but not before more than a dozen Adirondack lakes were reclaimed with it. At high concentrations toxaphene can cause nerve damage, immunosuppression, and organ damage, and the EPA classifies it as a probable carcinogen. Like the atomic elements of life, it enters food webs and is repeatedly passed from prey to predator, building up in the fat and muscle of consumers over multiple meals. Such biomagnification sickened fish-eating eagles and ospreys at the tops of their aquatic food chains before DDT use in the United States was similarly banned during the 1970s.

A former DEC employee told me that toxaphene was once used in Bear Pond, as well. "We loved it back then," he said, "because it killed the hell out of everything." However, the trout that were stocked in Bear Pond the following year died, and the same thing happened the year afterward. "They just wouldn't take," he said. Bear Pond's acidity might have been a factor, but reexamination of the reclamation procedure also suggested that several times more than the recommended dose of toxaphene might have been used accidentally.

I later spoke by telephone with Canadian scientist Brenda Miskimmin, who studied toxaphene in sediment cores. She told me that some fish in western Canada were still unsafe to eat several decades after toxaphene treatments, not only in lakes that had been reclaimed with the volatile compound but even in some lakes that simply lay *downwind* of reclaimed waters and farm fields where it was used as a pesticide. In her opinion, lakes that were exposed to toxaphene should be examined carefully to make sure that their fish are fit for human consumption.

I asked a DEC staff member if fish from the toxaphene-treated lakes in the Adirondacks were checked for contamination before people were allowed to catch and eat them. He said they were not because such contamination was unlikely and testing seemed unnecessary. I then sent a letter of concern to the regional director of the DEC, who granted permission to collect a trout from Black Pond and send it to a DEC lab for testing. The results came back negative, but the story of toxaphene in the Adirondacks is still unfolding.

In 2015, researchers at Clarkson University discovered toxaphene in a core that my students and I collected from Bear Pond half a century after the reclamation. It has apparently leaked upward into the most recent sediment layers and might therefore be circulating in the water and food web, as well. To my knowledge, no further studies on toxaphene contamination have yet been conducted in the region.

A more public disagreement erupted when another research project reopened the unresolved story of eutrophication in Black Pond. The idea for the study came from students in my paleoecology class. I had just finished explaining that the Black Pond cores revealed the timing of the eutrophication but could not identify its cause, because we did not know when the perch first entered the lake. A student then asked, "Why not look for perch DNA in the sediments?" I raised the issue with genetics professor Lee Ann Sporn down the hall, and she agreed to examine a trial core with us, using the lake next to our campus as a convenient study site.

On a frigid day in February, Lee Ann and I trudged over the windblown ice of Lower Saint Regis Lake with several students, drilled a hole, lowered a core sampler through it, and retrieved a cylinder of mud from the lake bed that was 53 inches (135 cm) long. Several weeks later Lee Ann came to my office with the results of her analyses. She and a student had found perch DNA from top to bottom in the core.

Lee Ann analyzed the DNA in a piece of perch fin and confirmed

Preparing to collect a sediment core through the ice on Lower Saint Regis Lake. *(photo by Curt Stager)*

the identification. She also analyzed a shorter core from the same lake and found perch genes in that mud as well. In order to be sure that the DNA did not come from fish molecules in the dust of the lab, she applied the same technique to sediment samples I had collected from a perch-free Adirondack lake, a fishless pool in the forest nearby, and Lake Tanganyika, East Africa. No sign of perch genes there. The DNA in the Lower Saint Regis core had come from the lake, not the lab.

Meanwhile, I analyzed the diatom record of the core and confirmed that the sediment layering was intact and that the DNA at the bottom of the core was therefore unlikely to have been stirred down from above. Carbon-14 results from a professional dating lab showed that the basal layer was two thousand years old. Together, our results told us that yellow perch are not recent immigrants after all, but have been native to this upland lake for thousands of years.

In hindsight, it seems unlikely that perch would not live in the Adirondacks. They are widely distributed in eastern North America and would presumably have been able to colonize upland lakes after the last ice age, just as the native trout did. Fred Mather, the author of the survey that found no perch during the 1880s, acknowledged in his report that his brief sampling of selected lakes was not definitive. Furthermore, the naturalist John Burroughs reported catching yellow perch in upland Lake Sanford and Nate Pond decades before Mather's study, and a map of the interior published during the 1870s included a telltale "Perch Lake" that Mather was apparently unaware of.

When our manuscript passed peer review at *PLOS ONE* and was published online in 2015, the public response was swift and intense. Blog posts and newspaper editorials spread the story widely, and an editor from the *Wall Street Journal* described the controversy along with the science in a front-page article. Opponents of reclamation mocked the DEC's assertion that the presence of yellow perch in a lake justified poisoning it. Some writers argued over the relative merits of perch-fishing and trout-fishing, and some disagreed over the ethics of fishing itself. Supporters of reclamation rejected our findings as environmentalist claptrap, and some DEC staff challenged our credibility as well as our results. One retired fisheries manager wrote a letter to the college president saying that he was disgusted with my "sloppy science" and that he would advise prospective students against attending our college.

Despite the controversy, questioning the management practices at Black Pond was important if only because it forced everyone involved to consider multiple perspectives on our relationships to lakes. Good intentions do not always lead to good policy, and science reminds us that lake communities are more complex than they may appear to be at first. Today more than ever, we cannot afford to let emotions and personal opinions blind us to the true inner workings of ecosystems, and our ways of interacting with lakes and the world at large must evolve as we learn more about them.

The story of Black Pond awaits further analyses that can support or refute the DNA results, but it nontheless offers potentially useful insights into the decline of native brook trout in the Adirondacks. Perch and trout often coexist elsewhere by dividing their shared lakes into separate territories. According to researchers in Nova Scotia, for example, perch tend to dominate the shallows there and leave the offshore waters to trout. The adults of each species also feed on the eggs and young of the other and thereby help to keep one another's numbers in check. If perch and trout shared Black Pond for millennia in similar fashion, then perhaps selective angling for trout during the nineteenth and twentieth centuries helped the less-desired perch to outcompete their neighbors and thereby invited the first reclamation. In that version of the story, people, not perch, are the problematic species who changed the ecosystem.

A local trout-fishing guide told me after learning of the paleo-DNA story, "People don't come to the Adirondacks to catch perch." That is not necessarily correct. If you want to guarantee success on a fishing trip with an impatient youngster, then a perch lake is an excellent destination, and Black Pond was a popular angling spot when it was loaded with perch. Many people also travel for ecotourism and value biodiversity enough for its own sake to want to protect it. If perch have lived here for as long as trout then they, too, might have developed unique heritage strains of their own. Such genetic diversity has been documented elsewhere in North American lakes, and local anglers will tell you that Adirondack perch can vary from lake to lake, including some whose black-banded flanks are light blue rather than yellow.

Classifying species as panfish or pests is understandable in terms of fisheries management but it overlooks a deeper reality, that lakes are also cradles of ecology and evolution with intrinsic value beyond food and sport. Genetically unique trout were probably lost during the first reclamation of Black Pond, not to mention possible rare strains of nongame species. In addition to their aesthetic and scientific worth, such locally adapted variants can have great practical value for management. Fish who have evolved in a wide selection of Adirondack lakes for thousands of years might contain genetic traits that could help their descendants to resist diseases, climate change, or other ecological challenges in an uncertain future.

As products of ecology and evolution ourselves, it is important to remember that we, too, are subject to their laws as we try to become better stewards of the Earth. In *A Sand County Almanac*, conservationist Aldo Leopold wrote of the need for a land ethic that "enlarges the boundaries of the community to include soils, waters, plants, and animals," and he urged us to learn how to "think like a mountain." Why not enlarge those boundaries to encompass a lake ethic, as well? After all, there are no lakes without land connected to them. And if controversial questions that are backed by good science also help more of us to think like a lake, then so much the better.

(photo by Curt Stager)

3

LAKES THROUGH THE LOOKING GLASS

Nature is to be found in her entirety nowhere more than in her smallest creatures.

—PLINY THE ELDER

THE DRY RATTLE OF dragonfly wings punctuates the soft hum of car tires on the edge of my old neighborhood pond. Tall elms and maples lean over a shallow pool no larger than the average yard here in suburban Manchester, Connecticut, where I spent most of my childhood. Their leafy canopy breaks the sunlight into fragments that rebound like sparks from the smooth, shaded surface, and the lush scents of uncut grasses and wildflowers mingle with that of damp mud.

For many of us, wild nature is seemingly found only in exotic, far-off locations, but I first discovered it in this humble backyard pond. It lies within a small glacial kettle hole on a vacant lot that is too uneven to put a house on, and most adults drive past it without a thought. But we local kids knew it well during the 1960s. We called it the Lily Pond, although no water lilies grew in it. We caught frogs on its banks in summer and skated on it in winter, but it became more than a playground for me at age twelve when a family friend gave me a used microscope that helped to launch my career as a student of lakes. Nearly five decades later, in 2015, I have returned to the pond that led me, like Lewis Carroll's Alice, through the looking glass into a world of wild nature in miniature.

Someone has placed a wooden bench beside the shore that serves me well now that age has made it more difficult to squat on my haunches as I used to. The Lily Pond is smaller than I remember, and not only because I have grown larger. Drought has downsized it to more of a puddle than a pond. Sedges and floating specks of green duckweed conceal much of the surface, but from this comfy perch I can still probe the water with my imagination. In my youth, I often dreamed of alpine adventures and African expeditions here. Now, after experiencing those things as a scientist, I better appreciate the secret pockets of untamed wilderness that can be found in unexpected places such as this.

Relatively small water bodies such as the Lily Pond are important habitats, in part, because there are so many of them. Recent analyses suggest that they are an order of magnitude more numerous than lakes the size of Walden Pond and roughly ten thousand times more abundant than the Lake Champlains, Tahoes, and Constances of the world. For the majority of their aquatic residents, as well, small size is only a relative concept. Most life on Earth is microscopic, and the world's most numerically abundant animals are tiny nematode worms. The arthropod phylum alone, which includes insects and crustaceans, contains far more species than all the world's mammal, bird, reptile, amphibian, and fish species combined. On those often-overlooked size scales, the puniest pond is a world of its own. With practice, we can learn to explore and enjoy those secret worlds, too.

Consider the dragonflies of the Lily Pond. Years ago, they were just "bugs" to me and one seemed much like another. I was wrong.

These dragonflies darting around me now spent most of their lives within the pond as wingless nymphs and only recently emerged as adults for a few final weeks of hunting, mating, and egg-laying. Most of them have long white abdomens like tapered cigarettes, the trademark of Common Whitetails. Four slim cellophane wings just behind the head

The Lily Pond. *(photo by Curt Stager)*

sport black and white bands that flicker like victory flags at a race car track. When gaps between clouds let the morning sunlight warm them up, the whitetails hover and zoom over the pond as they basket-scoop midges in midair with their slender, jointed legs.

Some people who know dragonflies better than I watch them through binoculars as bird-lovers watch birds, and each species has distinctive habits and features that may also be reflected in evocative names. Experts describe whole categories of skimmers, cruisers, spiketails, and petaltails, and within each category are many variants with names like comet darner, Stygian shadowdragon, and unicorn clubtail. More than five thousand species of dragonfly share the planet with us, and a hundred or so may live within this town alone.

Cruising among these whitetails are daintier creatures who belong to an insect family of their own, not of flying dragons but of fair damsels. Damselflies flutter more primly on their slimmer wings and they usually hold them upright at rest rather than horizontally. Many are decked out in iridescent greens, electric blues, and carbon blacks, suitable attire for the likes of sparkling jewelwings, powdered dancers, and sedge sprites.

I soon begin to notice more subtle differences from my vantage point on the bench. Some of the blue damsels are more brightly colored than others, and one oddball resembles a flying pretzel. The loop is actually two individuals in one, a mating pair. The bluer of the two is a male who has grasped a female by the nape of her neck with a clasper on the tip of his long, slender abdomen. She curls her own abdomen forward to receive a packet of sperm from her beau's underside and seems content to let him ferry her around thus. Such is the way of the Odonata, an ancient order of insects who were already flying over Paleozoic ponds when our first amphibian ancestors crawled out of the water. Mating for them is a drawn-out affair in which the lucky guy must guard his gal from other suitors who might take his place in the mating wheel, scoop his sperm packet away, and replace it with their own.

While watching the whitetails from this bench, I wonder what the world looks like to a dragonfly. Up close, the face resembles the bubble cockpit of a helicopter with two bulbous compound eyes that meet in the middle of the forehead. Each hemisphere contains thousands of gleaming facets that aim in all directions, and each facet is an eye in its own right. The dark shafts of the facets that happen to face you from any given direction produce the illusion of a pupil that seems to follow your movements.

For a dragonfly the world is a composite of many images like a wrap-around capsule packed with video screens. Snicker at a tiny bug-brain if you will, but first try to imagine seeing in all directions at once without becoming dizzy. Now imagine also navigating in three dimensions while flying at 30 miles per hour (50 kph) to chase tiny prey who can flee up, down, or sideways, all the while managing four flexible wings that move independently of one another. When it comes to brains, at least, size isn't everything.

It can be difficult to avoid thinking of insects in impersonal terms. We judge personhood by words and facial expressions, and odonates have neither. Their wings are their only sound-makers and their faces, like the rest of their bodies, are stiff and expressionless with chitinous armor. Many of us deny selfhood to insects for these and other reasons, but behavioral variation among individuals is raw material for evolution, and feelings such as fear, anger, pleasure, or desire are physiological traits that can have adaptive value. I am sure that I see signs of emotion and consciousness in these dragons and damsels, and I learn more about them by watching them as sentient individuals.

One of the whitetails returns over and over again to the same perch on a half-submerged grass stalk. After I watch his forays for several minutes I notice that he chases midges on some flights and trespassing dragons on others. This fellow (his behavior betrays his sex) is defending a territory. He knows that the slender green perch is his, and most of the passersby seem to know it, too, because they usually avoid him when he approaches. For those who need further convincing, he smacks into them to deliver a firm "buzz-off." His body twitches with tense energy before he launches, and sometimes he lifts his flashy abdomen erect like a warning flag. I suspect that every dragonfly on this pond knows and watches every other one as well as potential prey and predators.

There is also much going on here that I can't see directly. Dragonflies are ecological bridges between air and water. Through them, the ecology of the pond is reflected in the ecology of the vegetation around it.

The females here at the Lily Pond will soon lay hundreds of eggs in the shallows and then die along with the males. Their nymph offspring

Dragonfly adult (left) and nymph (right). *(photos by Curt Stager)*

will resemble miniature Humvee trucks and carry the fleshy pads of future wings on their shoulders. Dragonfly nymphs live submerged for months to years, depending on the species, and with the aid of a prehensile lower lip that can shoot forward like a fang-tipped arm they feed on smaller creatures of the pond. When threatened or seeking a meal, they squirt water from their rear ends for a jet-propelled lunge or leap. Each will eventually crawl out of the water, split a crusty exoskeleton down the back, and emerge as a winged angel of death who grazes down the local pollinators.

Dragonfly nymphs are fish prey as well as predators, but in fish-free waters they often survive to adulthood in larger numbers. Ecologists have recently demonstrated that insect-pollinated wildflowers living near fishless pools such as the Lily Pond often produce fewer seeds over the course of a growing season, all because of the dragonflies who breed and hunt there. I suppose the pollen-laden asters beside this bench might curse the bug-munching whitetails if they could.

I rise and walk to the water's edge, feeling the ground soften underfoot where the water meets the shore. Half-developed froglets bounce away from me like brown popcorn, splashing and vanishing among the reflected trees and clouds. I hunch down with more ease than I expected and gently sweep my fingers through the water as I often did decades ago, imagining what lies beneath the surface farther out in the pond. I am following in illustrious footsteps by doing so. During the nineteenth century the founder of my profession also found his career path as a kid who loved his hometown lake and the life within it.

THOREAU HAS BEEN called one of the world's first lake scientists because of his studies on Walden Pond, but the title more justifiably belongs to a Swiss physician and naturalist who was still a child when Thoreau began his lakeside retreat in 1845. François Alfonse Forel, or "FAF" as friends and neighbors called him, delved more deeply into limnology and also coined the term itself (Greek *limne* = lake, *logos* = word or discourse).

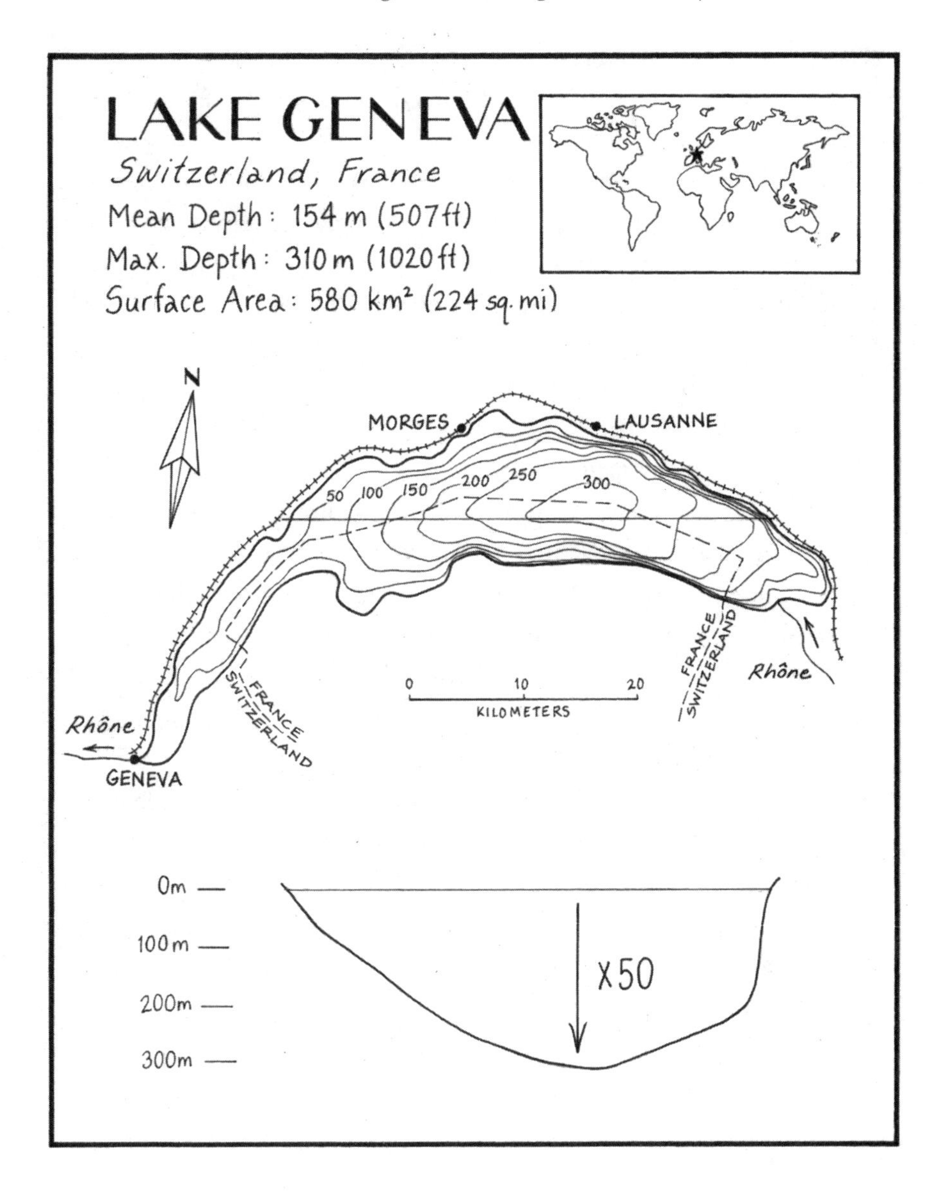

Forel grew up on the shores of Lake Geneva at the foot of the Alps in Morges, Switzerland, where he learned what he would later call "the art of observing and interrogating Nature." His childhood inspiration was quite different from my Lily Pond, which is small, shallow, murky, and lacking an outlet, features that most of us associate with ponds. Lake Geneva is 45 miles long (73 km), up to 1,020 feet deep (310 m), extremely clear, and drained by a major river (the Rhone), a classic "lake" rather than a "pond" despite the lack of a formal distinction between those two terms among limnologists.

Forel's youthful connections to his local lake gradually matured from boating and swimming with family into an all-consuming curiosity about what lay beneath the surface. As a teenager he impressed a team of archaeologists who were investigating the remains of a Bronze Age stilt village that projected out from shore by paddling his canoe among the ruins and hand-dredging three ancient bracelets from the bottom.

At age twenty-nine, Forel abandoned medicine as a career and accepted a professorship at the Academy of Lausanne near Morges. For the rest of his life Lake Geneva was his laboratory and aquarium, as he put it, as well as his passion. So intense was his enthusiasm for the lake that he initially refused a publisher's request to write a limnology textbook because he feared that he was too emotional about the topic to maintain his composure as a scientist. He later wrote the book after all but added this disclaimer to the foreword: "My relation to limnology is much too personal and subjective for me to be able to give an objective presentation of the facts." By the end of his life in 1912, he had learned more about lakes than anyone before him and established a new branch of science.

Forel made an accidental discovery during his first year at Lausanne that transformed his understanding of life in fresh water. While collecting sediments from the floor of the lake near Morges, he hauled a metal sampler-plate aboard and found it covered with fine silt. Rather than wipe it clean, he placed some of the muck under a microscope. He was astounded by what he saw. There among the flecks of sediment was a transparent, wriggling nematode. The bottom of the lake was not a sterile desert as he and other scientists had imagined, but very much alive.

To Forel, discovering that little worm was like finding life on Mars. Nobody had ever examined the cold, dark floor of Lake Geneva closely enough to recognize it as a habitat for strange and fascinating life forms, much less taken a microscope to it. The next day he dredged the bottom for more specimens, collecting small, mysterious creatures from deep basins and sunlit shallows. Even the water itself was teeming with life. "No serious analyses," he later wrote, "have indicated any lake water completely free of microbes." That was not a bad thing, he also noted. "All microbes are not necessarily unhealthy. Much to the contrary, the immense majority of these minute beings are completely innocent."

Instead of hoarding data and specimens for himself, Forel shared them among specialists in every field of study from physics, chemistry, and biology to the arts and human history. His new science of limnology was the study of all organisms and processes in and around lakes during the past, present, and future, including people; in essence, the study of everything.

You don't have to be an expert, travel the world, or own fancy equipment in order to explore aquatic wilderness and seek little-known species in much the same manner as Forel. I did it as a twelve-year-old with a field guide, a glass jar, and an inexpensive aquarium that I kept on a table in my bedroom. Like Forel, I soon discovered that there is much more to life beneath the surface of a water body than fish alone. Here are some examples of what turned up in my jar samples from the Lily Pond.

Discovery 1: tail-breather. A buff-colored beast as long as my forefinger hung head-down in my aquarium like a stick with six legs. A slim breathing tube the size of a toothpick connected the rear end to the surface. My field guide offered the sinister-sounding label of "water scorpion," presumably because of the toothpick, but my find was actually a hemipteran bug in the genus *Ranatra*. Water scorpions capture insect nymphs, tadpoles, and other small aquatic creatures with forelimbs similar to those of a praying mantis that swing open and closed like folding knives. Had he or she wished to, my temporary captive could have left the aquarium and flown back to the pond on paper-thin wings that were kept tucked away beneath leathery back-flaps. I would later learn that tail-breathing is not as unusual as I thought. Most other insects typically respire through pores in their abdomens, too, albeit without the tubular straw.

Discovery 2: mini-monsters. One morning the side of my aquarium facing the window was festooned with what appeared to be frayed bits of

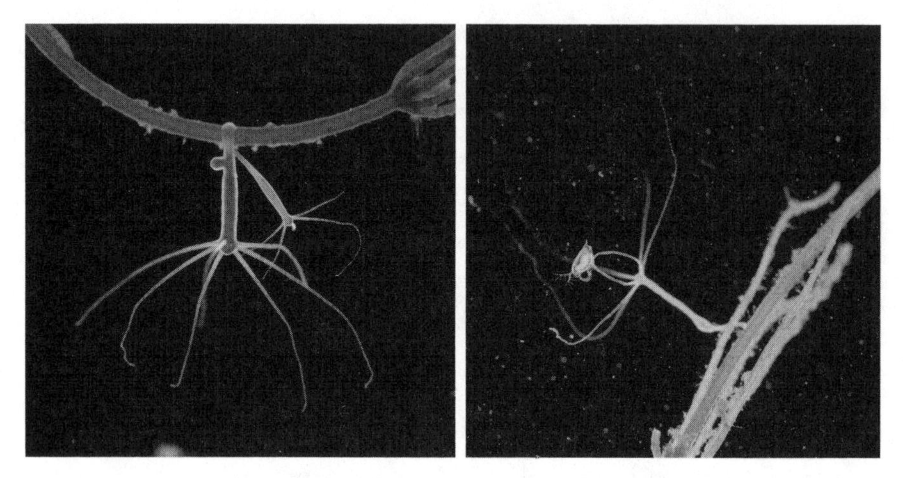

Hydras. Left: Adult with two offspring budding from the main stem. Right: Hydra capturing a cladoceran "waterflea." *(photos by Mark Warren)*

green string. They were hydras, miniature bottom-hugging relatives of jellyfish named after a beast from Greek mythology whose head was a tangle of venomous snakes. Each hydra dangled half a dozen stinging tentacles like willow wands and glided slowly along the glass like a walking tree. When a water flea blundered into one cluster of tentacles, the victim twitched, died, and was pushed down into the hollow body cavity to be digested. My field guide explained that green algae embedded in the body walls traded oxygen and sugars for the hydra's metabolic wastes. It was perhaps for the sake of their sun-loving algae that the hydras clustered nearest the window.

Discovery 3: gobbling feet. My pet snails, whose coiled shells resembled ram's horns, were not simply crawling about on the walls of the aquarium for fun. Through the glass I could watch a tiny round mouth pulsate in the center of each muscular foot. The snails pressed their mouth-feet against the algae-coated surface in order to scrape dinner with toothy ribbons that resembled the rasping tongues of cats.

When I realized that the brown slime I casually scrubbed away from the glass was food for the snails, I turned to my microscope for a closer look. Under the lens a smear of aquarium goo became diamond jewel boxes stuffed with amber chloroplasts, the glass-shelled diatoms I would later base my research on. Each algal cell was so small that fifty of them stacked one atop the other would barely match the thickness of my fingernail. The field guide told me that diatoms trap sunlight and produce oxygen just as plants do, and with the aid of the microscope I also learned that some of them could crawl like animals. My first impression of the brown coating had been as misguided as if I had looked down from an airliner and considered scraping the thin crusts of cities and forests away in order to polish up the landscape. I was hooked.

Through the microscope, hairy-looking pond scum that once disgusted me became jungles of transparent bamboo packed with light-harvesting emerald spirals. The so-called "water fleas" who fed the hydras were not fleas at all but fat, round cladoceran *Daphnia* and bullet-shaped, one-eyed *Cyclops* copepods. Deeper still in that microcosmos I glimpsed things even farther beyond the reach of normal vision, a little-known plane of existence that only high-tech scientific gadgetry and well-informed imaginations can detect. Floating among the algae and copepods were countless smaller dots, the faintly glistening cells of bacteria. They trembled like dim starlight in the background, so minuscule that they shivered from collisions with the invisible water molecules around them.

Calculations inspired by this ceaseless dance of particles, called Brownian motion, helped Albert Einstein and other physicists to confirm the existence of atoms decades before they could be viewed with powerful scanning tunneling microscopes as they are today. The same dance churns our water-filled cells, as well. A biomedical researcher once showed me how he used Brownian motion to distinguish living human cells from dead ones. He added bright green dye to a slurry of cells on a glass slide and placed the sample under a video microscope. We watched as the dye stained some of the cells and revealed closely packed flecks jiggling inside them. Those must be the live ones, I guessed wrongly. In fact, the activity arose from the Brownian motion of particles buffeted by water molecules and was not a sign of life at all. The living cells remained colorless because they actively rejected the dye, thereby concealing their own inner churn. Strangely, the dead cells looked more alive as a result.

Mini- and micro-sized life forms have always comprised the majority of species and, despite today's many environmental problems, great diversity and numbers of them still persist, more or less safely hidden in the realm of the very small.

LAKE GENEVA was a seemingly featureless gray plain when I first rode past it on a train to Zermatt on a cloudy day in January 1985. Although I had just completed a doctoral degree at Duke University that was based on the study of lakes, I knew nothing of Forel and the birth of my chosen field of study in Lake Geneva. My attention was instead focused on the rugged Alps where graduate student George Kling and I would soon be skiing beneath the Matterhorn. We had been delayed in Europe en route to joining our advisor Dan Livingstone on an expedition to the crater lakes of Cameroon, West Africa, and we were ill-equipped for an impromptu winter vacation.

The people in the seats around us wore expensive, trendy skiwear. We had only the light clothing we had packed for months of sweaty work in the tropics. When we weren't staring at the mountains we brainstormed ways to ski without shame or frostbite. The shameless part was easy: we both cared little for fashion statements we couldn't afford. Our solution to the cold issue was to wrap mosquito-proof bed-nets around ourselves to insulate our torsos and wear multiple layers of socks on our hands for gloves.

In hindsight, I know that our oblivious trip past Lake Geneva illustrated a problem that we all share. The world and its history are too vast and complex to grasp completely, and much of it is hidden beyond the range of easy perception. It is therefore wise to remember that a well-informed mind can help us to better understand what our senses report. Nowhere is that truth more apparent than on the scale of atoms and molecules. At that level of existence, the water we interact with in normal daily life becomes very strange indeed.

Forel and Thoreau documented the layering of water temperatures in Lake Geneva and Walden Pond, and they both noted how important it is for fish and other creatures. You can experience it yourself by swimming in a lake that is just beginning to warm from the top downward in early summer. If you kick your feet deep enough you might feel a sudden chill in your toes when they nick the cooler layer that fills the lower portion of the lake, what limnologists call the hypolimnion (Greek *hypo* = below) in order to distinguish it from the warmer epilimnion (*epi* = surface) on top. It wasn't until the last century, however, that we could also explain the layering of lakes on the molecular level. It arises from a simple principle that explains other important phenomena as well: water molecules are sticky.

You can observe the effect of that stickiness on small aquatic animals if you scoop a jar of water from just above the bottom of a shallow pond or a sheltered cove in a larger lake. Hold your catch up against the sky for backlighting and examine it closely. When the swirl of loose debris slows down, you may notice tiny specks that cross or resist the drift of the current rather than riding it. With luck, you will have caught some copepods.

Notice the jerky manner in which they move rather than easing along as smoothly as fish. That is an effect of trying to force one's way through hordes of sticky particles. Each short hop is like a leap into thick molas-

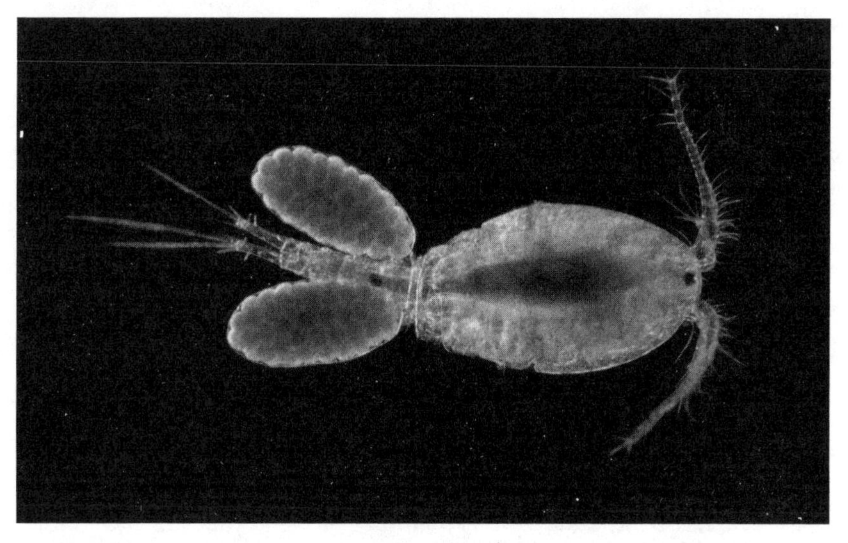

Female copepod with two egg sacs. *(photo by Mark Warren)*

ses that halts progress until the next leap, because the combined attractions between water molecules can stop microscopic creatures in their tracks. Imagine the turbulent chaos copepods must endure in that dense, viscous world when we swim among them.

If you watch copepods for long enough, you might also see another sign of sticky molecules. Sometimes, a passing individual will swerve to follow another one who swam past the same location a short time earlier. On that size scale the gluey liquid folds in around a scent plume and holds it intact until currents disperse it. In calm water, a female copepod can therefore trail a long filament of telltale pheromones behind her that is irresistible perfume to a potential Romeo.

More signs of elemental particles at work appear at the surface. The water molecules have nothing above them to stick to there, so they cling more tightly to one another instead. The springy, reflective film that results is several molecules thick and strong enough to support small objects. Look closely at the calm surface of a lake or home aquarium and you may see minute specks of dust or other detritus pasted to it like flies on flypaper. If you blow very gently or let a light breeze do the job, you will see the transparent sheet and its contents slide freely atop the deeper layers.

A six-legged springtail the size of a pepper grain can bounce on that stiff trampoline without denting it, but a larger water strider bug skims over it with dimples underfoot. Water striders skate on the pliant film in search of fallen insects to seize and feed upon, but they also use it to transmit messages to one another in the form of vibrations. Biologists at Cornell University once strapped micro-magnets to a female water strider's feet and electronically induced the tap-tap pattern of a strider love call in a lab tank. When the ripples from her unwitting invitation reached the sensitive feet of potential mates who shared the tank with her, the formerly disinterested males leaped to her side in response.

Forel would have loved to know enough about water molecules to weave them into his all-inclusive view of lakes and life. To do so without too much technical detail, we can simply imagine them as little blobs of matter with paired electromagnetic ears that provide the stickiness. The central blob of a water molecule is a single atom of oxygen and the two ears are hydrogen atoms. When nestled one against the other, the ears of each water molecule form weak hydrogen bonds with the oxygen chins of its neighbors. The blobs move with a chronic, random twitching that resists the pull of the hydrogen bonds and also drives the Brownian jiggling of larger particles that drift among them. Even when you sleep or sit as still as possible your molecular self is a seething mass of dancing particles, and so is every lake.

What do sticky bouncing blobs have to do with a swimmer's chilled toes or the layering of a lake? Heat drives the perpetual dance of matter, the warmer the faster. In summer, molecules in the cool hypolimnion slow-dance with their attractive hydrogen ears snuggled close to the chins around them. Meanwhile, the sun's heat goads molecules of the epilimnion into more frenzied motion that makes them crash against one another more forcefully and spread farther apart. The warm upper dance hall is now less densely packed than the cooler, more crowded basement, so it is lighter and floats on top.

These tiny particles are not only linked to one another but also to some of the world's largest phenomena, as when climates call the tune for the heat-driven dance of molecules in a lake. Forel described seasonal changes in the layering of Lake Geneva in units of temperature and density, but we can also do so in molecular terms.

In many lakes of the temperate zone, autumn cooling slows the dance of molecules near the surface so they pack more tightly and sink. Their sinking erases the summer strata, stirs dissolved oxygen to the bottom, and washes phosphorus and other algae-stimulating nutrients up from the mud. Similar mixing also happens after the spring ice-out for slightly different reasons. That seasonal mixing cycle is a lake's rototiller, fertilizer, and oxygen pump combined, and it drives the life cycles of many of its inhabitants.

Forel learned to follow the seasonal cycle in Lake Geneva by watching the color of the water change with the plankton. Like an artist he designed a standard color palette to document the changes throughout the year by comparing samples to a suite of blue, green, yellow, and brown dyes in numbered vials. More recently, scientists who monitor water quality in the same lake have lowered Secchi discs into it in order

Student using a Secchi disc to measure water clarity. *(photo by Curt Stager)*

to measure its color and clarity. Named for the nineteenth-century priest and oceanographer Angelo Secchi, the discs can be white or divided into black and white quadrants. A Secchi disc is lowered until it vanishes from sight, then pulled back up until it reappears at the so-called "Secchi depth," which represents a standard measure of clarity.

Clearer water produces deeper Secchi depths, and the white of the disc also reveals the color of the water while submerged. At Lake Geneva the Secchi depth usually shortens during the spring mixing season and again in summer when green and golden algae cloud the water. The intervening clearer phase marks the birth of copepods and other phytoplankton-eaters who graze the first bloom down and then die back enough for the algal community to recover.

Lake Geneva doesn't freeze over in winter because the local climate is too mild and the lake is too deep. For those lakes that do, the effects of the sticky molecules can seem to challenge common sense. Why, for example, does ice float? The tightly interlocked molecules of most solids are more densely packed than those of liquids, so one might expect solid water to sink like a stone and build up on the bottom of a lake. Fortunately for fish and other aquatic life, it makes a protective lid instead.

The lid forms because the lake's water molecules chill and slow down as winter approaches. Rather than crash and bounce apart from one another, they begin to link up and produce six-sided rings with an empty space in the center of each. The rings then connect to their neighbors until the surface of the lake puffs up into a stiff, buoyant lattice, the molecular Styrofoam we call ice. We are lucky that most of the water molecules we encounter jiggle as much as they do. Were it not for the goading of the Sun and the heat-trapping atmosphere, Earth's water would be just another solid mineral as it is among the frigid outer planets of the solar system.

Fish and turtles survive just fine under winter ice as long as their lake contains sufficient dissolved oxygen from the autumn mixing period. When new ice is still transparent but thick enough to support a person safely (4 inches or 10 centimeters, at least), you can sometimes spot fish in the shallows beneath you and chase them around on foot. Snapping turtles normally burrow into the muddy bottom and slow their metabolisms enough to survive the winter on oxygen that diffuses through the linings of their mouths and throats. I have also heard of snapping turtles being seen through clear ice as they patrol the floors of northern lakes, but I have not yet heard an explanation of how they might survive an active lifestyle through an entire winter without coming up for air to power their oxygen-hungry muscles.

The secret strangeness of water reorganizes the whole lake again in spring when sunlight grows stronger and lasts longer. The lattices within the ice tremble more and more vigorously until they collapse into broken rubble that flows as a liquid at 32°F (0°C). However, some bulky pieces of the melted lattice remain intact until the waxing Sun breaks them down further into single molecules that pack more closely together. For this reason, the water actually becomes more uniformly compact and therefore denser as it warms until it reaches maximum density at 39°F (4°C). In the process, all layering is lost and the lake mixes once again until more surface heating builds a new epilimnion.

In sum, the warming and cooling of sticky, dancing water molecules drives the seasonal cycles of most lakes in the temperate zone. The bottom water is densest and heaviest in winter because, oddly enough, it is slightly *less* cold than the nearly frozen water just under the ice. The coldest-on-top layering in winter must therefore reverse in order to form the coldest-on-the-bottom layering of summer, and vice versa. Limnologists say that a lake "turns over" during those thermal flip-flops of spring and autumn. Twice-annual overturns are typical of temperate lakes, but the pattern varies with depth, location, and other factors. The shallow Lily Ponds of the world, for instance, often lack a hypolimnion altogether and mix whenever a brisk wind stirs them, but many deep, stratified lakes of the tropics mix only once a year if at all.

Our prescientific ancestors could only invent mythic or magical explanations for such things if they even noticed them at all, and replacing those stories with the invisible forces and phenomena that science reveals may seem like an arbitrary choice. It is not. This is a fundamental threshold of awareness that we are crossing with the aid of modern science, because molecules and the processes of life that they produce are demonstrably real. Learning about the sticky molecules within a lake is like adding subtitles in our native language to a foreign film. When we know more of what is really going on we can better understand and savor the story instead of simply guessing at it.

One of the most pleasant ways to learn more about life in water is through an immersion experience in which one becomes part of it, and I remember doing just that in a deep blue crater lake in Cameroon in 1985. It happened several weeks after George and I left Lake Geneva for the kind of African expedition I had imagined at the Lily Pond as a child.

BAROMBI MBO OCCUPIES a circular volcanic explosion crater that is 1.5 miles wide (2.5 km), 360 feet deep (110 m), and rimmed with lush tropical rainforest. Access to it from the adjacent town of Kumba is by a steep dirt track, and a small fishing village in a clearing on the opposite

Barombi Mbo, Cameroon. *(photo by Curt Stager)*

shore is reachable only by dugout canoe. The lake's name is pronounced "bah-ROM-bee mm-BOH" and translates to "lake of the Barombis" in reference to the villagers who claim it as their own.

Clouds and seasonal rains cool the epilimnion of Barombi Mbo and allow some of it to sink from July through September, but not enough to fully recharge the lower 200 feet (60 m) or so with atmospheric oxygen. In most temperate zone lakes that mix regularly, the hypolimnion is an oxygen-rich refuge for trout or other cold-loving species in summer, but in permanently stratified Barombi Mbo it is a no-go zone for most animals. Nonetheless, members of one of the eleven species of cichlid fish (pronounced SICK-lid) endemic to the lake dart down into it to capture midge nymphs who hide there. Extra hemoglobin in their blood helps the deep-diving cichlids to absorb what little oxygen is available, and local fishermen report that the little fish bleed from their gills when pulled to the surface.

The lake was calm and comfortably warm on that remembered day in 1985. The luxuriant canopied forest was a fragrant greenhouse, and a

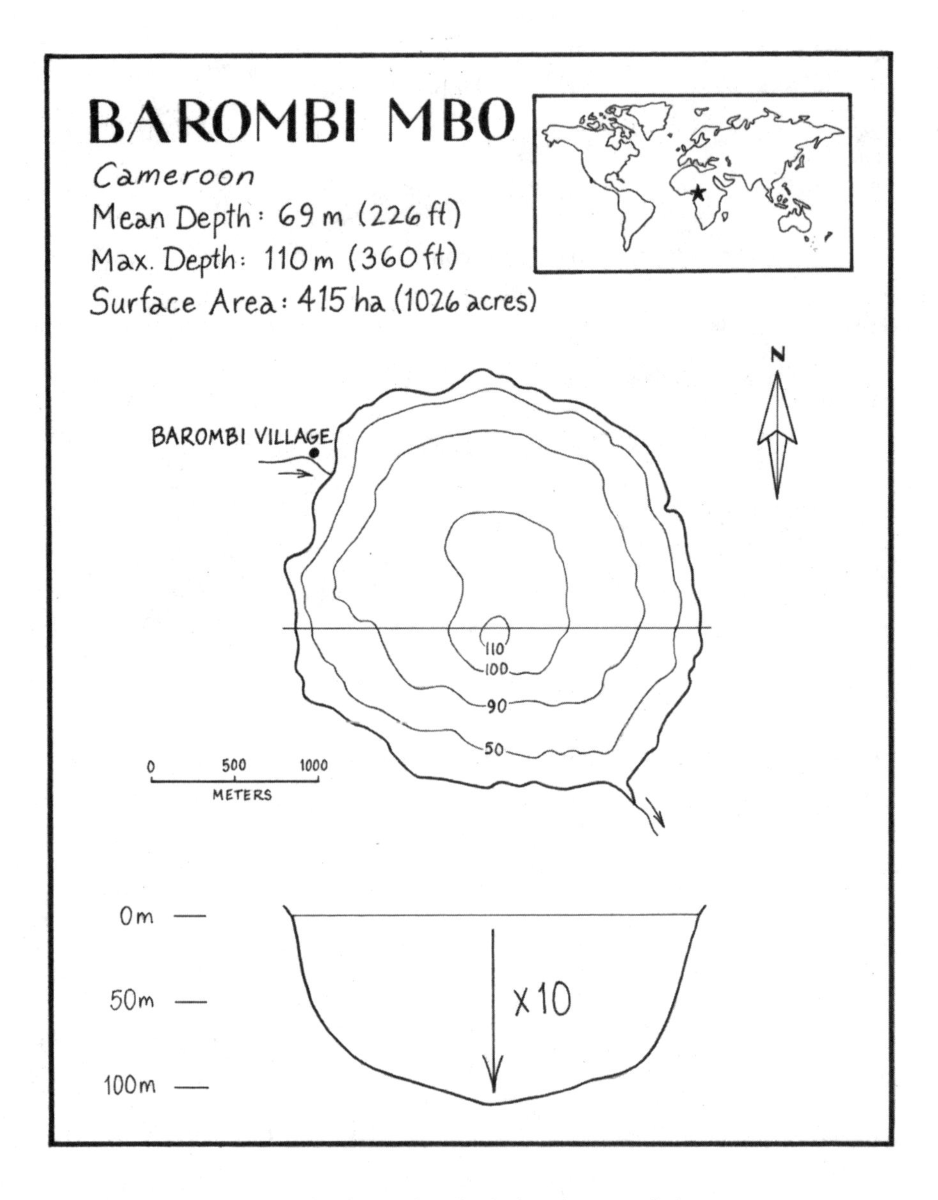

troop of chimpanzees had just finished shrieking and drumming against resonant tree trunks on the far side of the crater. The air was so hot and thick with humidity that the slightest movement soaked me in sweat, all the more incentive for a swim. My attention at the moment, however, was focused on gaining a better sense of what life inside the lake felt like.

I climbed atop a large, sun-warmed boulder beside the water's edge and dove in. Immediately, my heart rate slowed, as did the circulation in my limbs. This was an involuntary reaction to water on the face, what physiologists call a "dive reflex." Nobody is sure why we have it, but many other mammals possess it as well. Seals and whales have more refined versions of the reflex that help to conserve oxygen at depth, as do extra hemoglobin and myoglobin in their muscles. In colder water, the human version of it might also come with an involuntary gasp that can lead to drowning.

I soon noticed my need for air, something I normally took for granted. If I were to try to breathe the water it would be too thick for my lungs to handle and too depleted in oxygen to sustain me. Oxygen represents about one-fifth of the molecules in air but well below 1 percent of those in a lake. Fish survive on dissolved oxygen because they have slower metabolisms and don't have to pull the water into and out of paper-thin air sacs as we do. Instead, they pump large amounts of it down their throats to wash over frilly, oxygen-absorbing gills, then release it through twin slits behind their heads to extract the oxygen they need.

I stayed under for a few more seconds and opened my eyes. Everything was a blur. Human eyes don't see well in lakes because our curved corneas aim waterborne images close to the center of the skull rather than on the retinas. For me to see as well as a fish down there, I would have had to look into the water through a layer of air as if looking into an aquarium. I returned to shore, donned a mask and snorkel, and pushed off again with my glass-plated face forward, arms extended, and feet fluttering.

The water was so transparent that my hands stood out starkly against the luminous blue background as though I were holding them up against the sky. As Forel noted, the color and clarity of lake water varies with the substances that wash into it and the kinds of organisms who live in it. Parts of Lake Geneva are milky with glacial silt from alpine rivers, and the Lily Pond is stained brown by leaf litter and often turbid with greenish plankton. Barombi Mbo's water was so clear because it had never known a glacier, the duff on the forest floor was quickly destroyed by microbes and absorbed by tree roots, and the plankton was widely dispersed and rapidly eaten.

Despite the clarity of the water up close its molecules also scattered light and prevented me from seeing objects much more than 30 feet (9 m) away, an aquatic version of the blue atmospheric haze that colors distant mountains. You can't look across a lake from below the water line, nor can you readily trace its submerged features with your eyes as you can a landscape. Vision in water is for close encounters, and most of it must be done near the sunlit surface where breeding colors are brighter and both predators and prey are easier to see. A lake's plants need light for photosynthesis too, as do rock-encrusting or free-floating algae, and most of the shelter and food that many fish seek are therefore found in shallow water.

A sharp thud reminded me that I was not alone. George was working on a large raft of wood and aluminum that we had moored in the center of the lake for sediment coring. Whatever equipment he had banged on the raft sent a shock wave into my ears. Sounds travel faster and better through liquid water than air because molecules are more tightly packed in liquids than in gases, and the sticky hydrogen bonds between water molecules pull them even closer together. When a noise jolts them they transmit it quickly by bumping into their neighbors with little space between them to buffer the motion.

Sound often works better than sight for communicating over long

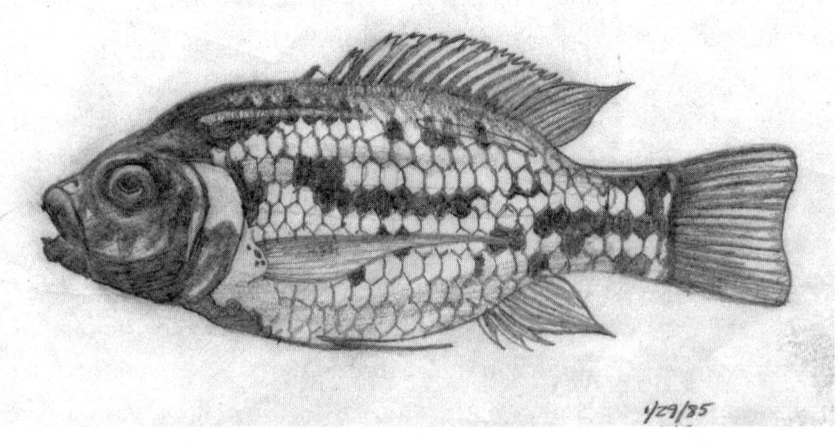

Pungu maclareni, a species of cichlid unique to Barombi Mbo.
(illustration by Curt Stager)

distances in a lake, and many aquatic creatures take advantage of it. I had recently been surprised to hear high-pitched rasping sounds while snorkeling in another crater lake north of Barombi Mbo. They were the courtship calls of water boatman bugs the size of sunflower seeds who hovered around me in the water, chirping like crickets. If I were a fish I would hear such things often, and I would also sense them along the full length of my body. Look closely at a fish and you can usually see a pencil-thin line running along each flank. Each lateral line is a tube full of fluid in which sensory hairs bend when shock waves strike them through pores in the tube. The high fidelity of waterborne sounds when heard with both flanks must make for quite a concert when a fish listens to a lake.

Before heading for the raft I held my breath and dove down about 10 feet (3 m). The water cooled around me as I left the extra-warm layer that had accumulated at the top of the epilimnion since morning and would disperse overnight. The hypolimnion lay much deeper in the blue mist below me, and if I had been able to reach it I would have felt more of a chill, some 4–5°F (2°C) cooler. Above me, the surface was a gently billowing silver canopy. I kicked forward and in the half-minute that my lungs allowed I savored a sensation that I recalled from dreams. I was not swimming but water-flying. The density of my fluid-rich body was similar to that of the invisibly clear medium around me and made me feel that I was suspended in midair.

Fifteen minutes later I joined George on the raft. With him was Ndoni Sangwa Paul, a fisherman from the Barombi village whom we had hired to help us with our fieldwork. Paul had the beard, build, and demeanor of Abraham Lincoln and had quickly become a friend as well as a collaborator. He also possessed a wicked sense of humor.

“Monkey work, Bambo chop,” he said with a wry smile when I climbed aboard, dripping wet. By that he meant I was a slacker. I had been acting like Bambo, the lazy chimp who spends his time eating, or in my case playing in the water, while he and George did all the work.

During our stay in Cameroon we would collect sediment cores that reached twenty-five thousand years into the past, almost twice as deep in time as the history of Walden Pond, in order to study how the resident cichlid fish evolved and how climate changes of the past affected the local rain forests. But even the core that would later prove to be one of our longest, hauled up in a string of steel pipes nearly 80 feet (23.5 m) long, would merely scratch the surface of Barombi Mbo's full million-year history.

We were not probing the past on that day, however. Instead, George and Paul were investigating a mysterious pulsation in the water beneath the raft. While monitoring a submerged temperature probe, George noticed that the water 100 feet (30 m) below us had repeatedly cooled and warmed in cycles lasting about eighty minutes. It meant that the top of the hypolimnion was rising and falling as though the lake were breathing.

Forel documented something similar on the surface of Lake Geneva under windy conditions like those that Barombi Mbo had experienced recently, a gradual oscillation of several inches akin to the sloshing of water in a bathtub. He noticed that currents flowed into and then out of the harbor at Morges on a regular basis and realized that the motion was caused by a slow, tide-like wave powered by wind. Local residents called it a "seiche" (pronounced "SAYsh"), a label that Forel also adopted. It wasn't until limnologists later studied the deep lochs of Scotland, however, that hidden seiches were discovered far beneath the surfaces of lakes, too.

Internal seiches often form when strong winds push water to the far side of a deep, stratified lake where the extra weight squashes the hypolimnion downward until the winds subside and let it seesaw back up. They can also develop beneath heavy surface swells. The similar densities of the upper and lower layers allow the resultant waves on the interface between them to grow much larger than those on the surface. The internal waves within Loch Ness, for example, rise and fall by more than 30 feet (9 m) and can measure more than half a mile (about one kilometer) from crest to crest. Such internal seiches may break into invisible surf during storms and sweep the shallow regions of a lake clean of sediment, phenomena that can be both a gift and a threat to life in the upper layers. The turbulence stirs nutritious elements up into the surface waters where light-loving algae can reach them, but it can also push oxygen-starved bottom water up to the surface and kill fish.

Paul said that he had never seen a fish-kill due to upwelling in Barombi Mbo, nor did he or anyone else know about the internal seiche until today. He did, however, know a lot of traditional lake-related lore, and he shared some of it with us on the raft during our lunch break.

According to Paul, an ancestor of the Barombis discovered the lake while hunting with his dog in the uncharted past. The lives of the villagers were closely tied to it. They drank it, bathed in it, washed clothing in it, and traveled across it in their canoes. The men netted its fish to sell in the market in Kumba. The women gathered gray lacustrine clay from a secret spot in the forest that water once covered before the outlet stream cut a notch in the crater rim and lowered the surface level of the lake. They used the former lake mud to make flask-shaped storage vessels that doubled as hand drums during celebrations and ceremonies.

The Barombis were connected to their lake not only as a resource but also on deeper personal levels, as well. The spirits of chiefs and other

Fishermen visiting our coring raft on Barombi Mbo, Cameroon.
(photo by Curt Stager)

notable forebears were said to live in a village on the bottom of the lake, and some of the Barombis worried that our sampling equipment would disturb their ancestors. In order to address their concerns, Dan and the rest of us had met with several elders in the chief's thatched house. They asked Dan to provide food and drink as an offering to the spirits, and he obliged on the condition that we could watch the ceremony.

When the anticipated day arrived most of the food and drink went to the villagers, but a cup of chicken blood was also poured into the water with a prayer. I would later encounter so many similar traditions while working in Africa that I became used to offering symbolic gifts to lake spirits before sampling in the field, a practice that I still maintain today.

I went to Cameroon to study the crater lakes, their sediments, and their fish, but after spending several months among the Barombis I began to think of people as integral members of lake ecosystems, too. Forel felt the same way about Lake Geneva and the people who lived near it. When he compiled a summary of species associated with his beloved lake, humans topped the list. "*Homo sapiens* . . . is not an essentially aquatic species," he wrote, "but has become so by way of his activities; the calling of fishermen, sailors, . . . etc. results in many such people living a semi-lacustrine life . . . [as] an erratic species of the lake fauna."

DURING MY PILGRIMAGE to the Lily Pond in 2015, half a dozen ducklings and their mother seem to share Forel's opinion of humankind's place in an aquatic community. They spot me from across the pond and rush over to my bench, only to waddle back to the water when they learn that I am a poorer food source than some of the current residents of the neighborhood. The place looks and feels much like it did in my youth, but it has also changed since I last saw it, as have I. The pond has become so shallow under the influence of drought and natural in-filling that I am tempted to concur with a neighbor who stops by to chat. He says that he wants to have it dredged with a backhoe before it fills in completely. I am not certain that such a move is necessary, but I am glad that people still care enough about this place to want to save it.

The Lily Ponds of the world are easy to overlook, hidden by their familiarity but even more so by their size, as are the tiny organisms that thrive within them. Nevertheless, they can introduce an attentive visitor to the secret world of lakes as surely as an exotic Geneva or Barombi Mbo. Like Forel and the Barombis, I will always feel a special affection for the still waters of home that first opened that world to me half a century ago.

(photo by Kary Johnson)

4

THE GREAT RIFT

A grain in the balance will determine which individual shall live and which shall die—which variety or species shall increase in number, and which shall decrease, or finally become extinct.

—CHARLES DARWIN, *On the Origin of Species*

IN NOVEMBER 1960, my future mentor at Duke University, Dan Livingstone, launched a small inflatable boat on a remote lake in Zambia, south-central Africa, with his graduate student, Joe Richardson. Then in his thirties, Livingstone was an expert in lake ecology and environmental history, and he and Richardson had come to Lake Cheshi, a swamp-rimmed pocket of open water connected to a much larger lake, Mweru Wantipa, where they planned to collect a sediment core. What they didn't plan on was a lesson in natural selection, courtesy of a very large, angry crocodile.

A day earlier, fishermen camping on the shore had welcomed the foreigners and shared with them their knowledge of the lake. This was not the first time someone with Dan's family name had stopped by unexpectedly. The Scottish explorer and missionary David Livingstone passed through this region in 1867. Four years later the Welsh-born

Joe Richardson (left) and Dan Livingstone (right) preparing to core from an inflatable boat in East Africa, 1960. *(photo by Robert Kendall)*

American adventurer Henry Morton Stanley found him at Lake Tanganyika and greeted him with the words, "Dr. Livingstone, I presume?"

The fishermen invited Livingstone and Richardson to launch their boat through a narrow channel that had been chopped through a quarter-mile of papyrus swamp, but they also warned them to beware of hippos. Sharp-fanged hippo parents were known to bite marauding crocodiles in half if they came too close to their calves, and watercraft were potential targets, too.

All seemed to go well at first. Livingstone and Richardson motored slowly across the lake in order to measure its dimensions and identify a good coring site. It was 8 miles (13 km) long and nearly 3 miles (4 km) wide, with a flat bottom of soft mud in water that was just barely over their heads, well suited for coring. Having finished their survey, they turned back toward the fishing camp.

They were still in the middle of the lake when a thunderstorm appeared on the horizon. The boat was just 6 feet (2 m) long with very little freeboard, hardly the kind of craft to weather a storm in. Then they ran aground on something. "I thought it was a log," Richardson recalled during a visit to my lab five decades later. "Only this log had teeth."

Suddenly, a gigantic crocodile burst the front compartment of the boat with a single bite and raked Richardson across the buttocks, bruising but not breaking the skin. The monster's gray-brown back was more than 1 yard (1 m) wide, and the partially deflated boat teetered on that writhing platform while Richardson tried desperately to push them free. Massive jaws clamped down on Livingstone's foot through the now-flabby floor lining until he kicked it loose. The croc then submerged, reappeared at Livingstone's end of the boat, and clambered up onto the intact stern as if to sink it. Richardson used the only weapon available to try to fend off their attacker—a hollow aluminum paddle that they carried in case of engine failure. As he later told me, "the blows had very little effect on the animal."

In the confusion, both men tumbled into the water and began to swim for shore. Fortunately, the crocodile was more focused on the limp rag of a boat than on them. Their last sight of the wreckage before waves blocked their view was of the croc thrashing about on it as though trying to drown it.

Now the water itself became a more immediate threat. The storm struck with full force and whipped the waves into foam. Livingstone was a poor swimmer, and if they were to reach shore they would have to cross a mile of wind-lashed lake. Richardson pushed Livingstone along while trying to keep their heads above water amid the waves. Two hours later, they reached the reed swamp that blocked their access to solid ground.

Fear of being eaten returned then because Lake Cheshi's crocodiles often hunted near the edges of the lake where baby hippos and fish were most plentiful, but luck was with them. They found the opening to the channel and finally made it ashore with little more damage than bruised buttocks, a puncture wound from a machete-cut papyrus stem, and lost equipment.

Livingstone and Richardson later burnished their reputations in the scientific community by describing their croc-encounter in an article for the journal *Copeia* titled, "An attack by a Nile crocodile on a small boat." In it, they blended meticulous scientific observations with understated drama, explaining that their attacker's muzzle was "even broader than the muzzles of crocodiles four and a half meters in length," and "the pupils of its eyes were closed to narrow slits." They also proposed a possible reason for the attack. Previous assaults on motor-powered boats on Mweru Wantipa suggested that the vibration of the engine sounded enough like the roar of a bull crocodile to summon a territorial response.

The horror of becoming prey, an experience that was well known to our distant ancestors, is now a feeling that most of us know only vicariously. "The Croc Story," as it later became known among Livingstone and Richardson's family and friends, is both a gripping tale of primal adventure and a page in a larger saga that plays out on longer time scales. Predators not only eat their prey but also shape the species that their prey belong to, as do territorial disputes, the quest for mates, and the shifting of climates. In this manner, many individual lives and deaths comprise a tapestry of evolution, the story of a species.

Evolution is not merely something that happened long ago, but an ongoing process that still operates today. Recent debates over the ethics of genetically modified organisms (GMOs) may suggest that mutants and composite life forms are unusual outside the laboratory, but that is not so. Our own genomes are collections of repeatedly transformed, multispecies traits that accumulated over billions of years and link us to all life on Earth. Our genetic code originated in bacteria. The hemoglobin in our veins is a gift from early marine invertebrates. The genes that shape our jaws arose first in fish, and the amniotic sac that helps to sustain a child in the womb is a former reptilian invention that also protects baby crocodiles inside their eggs. What makes synthetic GMO mutants most unusual today is not their altered nature but the conscious thought processes that direct their evolution.

Even without artificial GMOs in the picture, evolution remains a dynamic and universal fact of life. Many of us overlook it because we normally focus on brief slices of the here and now rather than broader perspectives that reveal evolution in action. Nonetheless, we humans are not only products of evolution but also increasingly powerful agents of it, as the lakes of Africa attest.

THE LAKE-STUDDED Great Rift Valley system of East Africa is more than 2,000 miles (3,200 km) long, an ancient network of tectonic cracks in the Earth's crust that may one day set the Horn of Africa adrift in the Indian Ocean. One arid eastern branch runs from the mouth of the Red Sea through Ethiopia and Kenya to the Serengeti Plains of Tanzania. Petrified bones of our early hominin relatives, including Lucy, Turkana Boy, and Zinj, have been unearthed there as well as those of even older forebears. A wetter western branch curves south to Malawi and contains some of the largest, deepest, and oldest lakes in the world. Two of them, Tanganyika and Malawi, fill 400-by-30-mile (640 by 50 km) stretches of it with as much as 4,820 vertical feet (1,470 m) of water, and are millions of years old. Where Kenya, Uganda, and Tanzania meet on a plateau between the two arms of the rift lies Victoria, the world's largest tropical lake and the main equatorial source of the Nile River.

The lakes of the Great Rift are diverse and spectacular. The shoreline of Lake Bogoria boils with geysers and bubbling hot springs. Tomato-red Lake Natron is so caustic with alkaline minerals from volcanic soils that it can destroy flesh. Shallow lakes Elmenteita and Nakuru often turn pink with thousands of wading flamingos who are themselves painted by the pigments of brine shrimp that their curved bills filter from the corrosive water. Lake Kivu is a potentially explosive cauldron 50 miles (80 km) long and more than 1,500 feet (480 m) deep that is supercharged with subterranean CO_2 and bacterial methane. Limnologists worry that a sudden landslide or volcanic eruption could unleash Kivu's gas on surrounding towns in a violent rush of flammable froth.

My first experiences with those remarkable lakes came in 1981 when I spent three months among them as a graduate student on a geophysical survey and coring operation that Livingstone had instigated and helped to organize. The sheer size of the largest lakes can be difficult to grasp. As we cruised high above Kenya on our way south to Malawi, a geologist in the seat next to me pointed to a round water body far below us and exclaimed, "There's Lake Victoria!" Not knowing better yet, I nodded. Then an enormous gray plain of water stretched to the horizon. The impostor was Lake Naivasha, 8 miles (13 km) in diameter. The immensity that lay before us was only one small sector of the real Victoria.

Lake Victoria is roughly 200 miles (320 km) long and 150 miles (240 km) wide, nearly as large as Ireland and second in area among freshwater bodies only to Lake Superior. Despite its tremendous size it is only about 260 feet (80 m) deep at the center, a gigantic puddle of backponded river water that formed when rift walls rose on either side of it. To view it in its entirety was impossible until we could look down at it from orbit. What even an astronaut cannot see from above, however, is another superlative feature. Evolution has worked overtime on the fishes of this lake, spawning hundreds of endemic species that are found nowhere else.

Colorful cichlids are famous among evolutionary biologists and ama-

"Mbuna" cichlid from Lake Malawi. *(photo by Curt Stager)*

teur aquarists alike. The fish family Cichlidae is found throughout the tropics and subtropics, and its enthusiastic fans tend to praise a given flock of species from a particular region like a home sports team. Among the African-cichlid crowd, however, the Victoria species are especially valued because many of them face possible extinction.

The rise and fall of the Victoria cichlids is an epic tale spanning thousands of years that meshes in surprising ways with our own history. It can be challenging to watch Victoria cichlids in the wild, however, because much of their home lake has recently lost its transparency to water quality problems. Lake Malawi, our first destination on the geophysical survey and an even richer showcase of species diversity, remains extremely clear and therefore more suitable for fish-watching. It supports as many as a thousand kinds of cichlid all its own, about 10 percent of the world's known freshwater fish species. To swim among them is like visiting a coral reef full of technicolor life.

When my colleagues and I arrived at Nkhata Bay on the northwestern shore of Lake Malawi in 1981, tropical trade winds plowed the surface into white-capped swells. Even the strongest gusts, however, couldn't stir airborne oxygen anywhere near the bottom, which lay almost half a mile (706 m) below the waves. Lake Malawi is two lakes in one, a sunlit zone of abundant life atop an oxygen-poor trench of death. Most of Lake Malawi's animal residents are restricted to a narrow ring of shoreline habitats within the uppermost 800 feet (250 m) or so of the steep-walled trough, with the notable exception of lake-fly midges.

Each year, more than a hundred thousand metric tons of transparent, thread-like *Chaoborus* nymphs develop deep beneath the surface of Lake Malawi. For months after hatching they hide by day in the dark, oxygen-starved depths where most fish cannot follow, and then migrate hundreds of feet upward by night to hunt zooplankton. When they are ready to mature into winged adults, air sacs pull them rapidly up through a gauntlet of hungry fish who eat three-quarters of them before they reach the surface. Those who survive erupt into mating swarms so thick that the water seems to smolder with living smoke.

The adult flies resemble mosquitoes but they lack biting mouthparts, and their sole aim is to breed and lay eggs before dying just a few hours

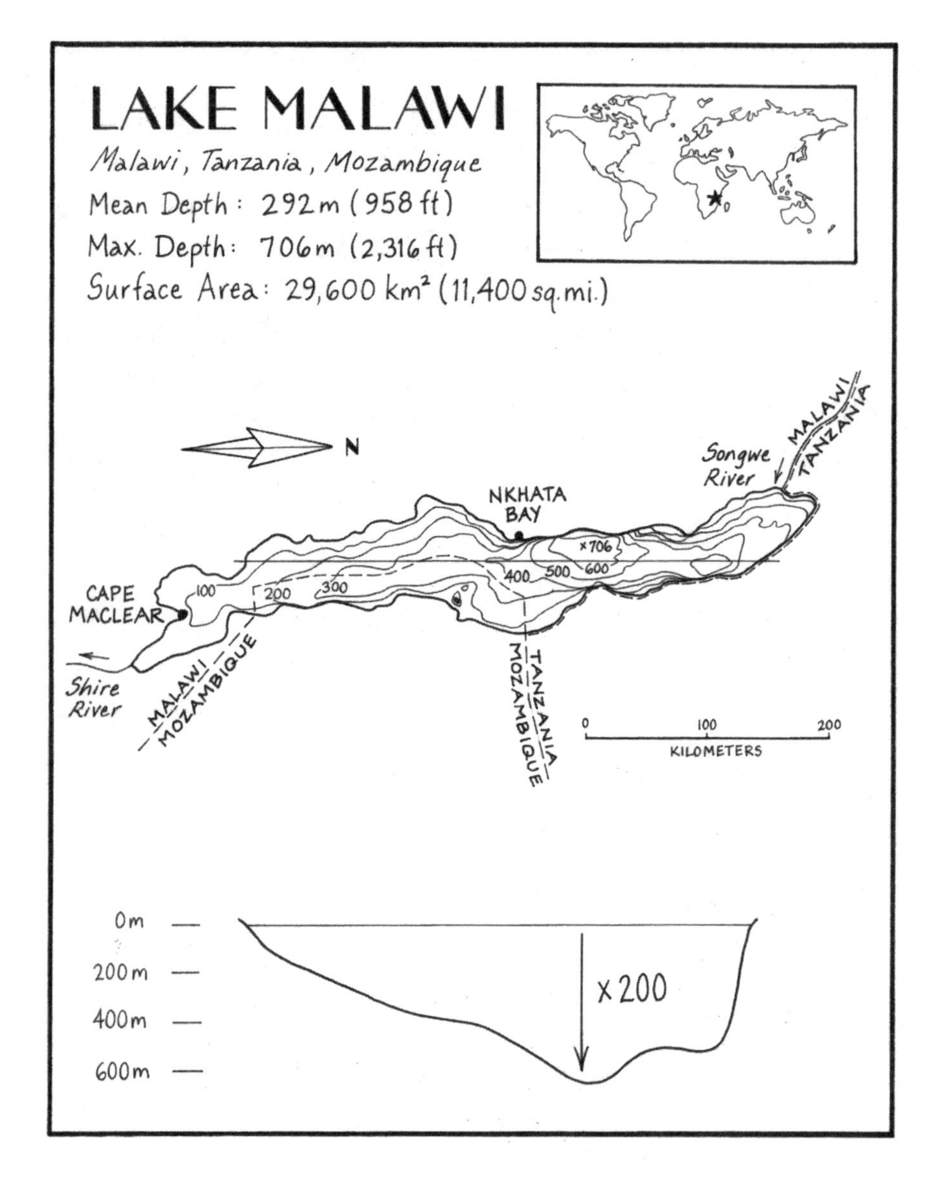

to days later. Harmless to humans, a lake-fly swarm is nonetheless disconcerting if your boat plows into it or it drifts ashore and engulfs you. Everyone and everything is suddenly covered with soft buggy fluff. Birds feast on the bounty and so do local people, who boil and crush the protein-rich insects into cakes, which is why the species carries the name *C. edulis* (meaning "edible" in Latin).

The flies of Lake Malawi illustrate classic natural selection by predators. Many more flies are born than can survive, and slight variations that are inherited at birth can determine their odds of survival. Flies who rise much faster or earlier than average are more easily spotted in isolation, as are the slowpokes. Fish therefore tend to preserve the present form of *C. edulis* in Lake Malawi by trimming extreme variants from the population.

The dynamic mechanisms that maintain the apparent stability of species such as these were summarized by evolutionary biologist Leigh van Valen in his "Red Queen hypothesis." Borrowing a mad character from Lewis Carroll's *Through the Looking Glass*, van Valen used her to illustrate how seemingly unchanging species continuously adapt to evolving predators, climates, and other challenges. "Now, here, you see," the Red Queen said to Alice, "it takes all the running you can do, to keep in the same place."

The flies of Lake Malawi remain recognizable from generation to generation because the selective pressures they face happen to keep them close to their current form. In this case, natural selection is a source of stability in the lake-fly gene pool despite a continuously bubbling spring of mutations, but it can also be an engine of evolutionary change. In the more distant past, the first super-buoyant mutants among early ancestors of today's flies might have replaced their compatriots because they were less likely to be eaten in a traditionally slow migration to the surface and therefore more likely to survive and pass the new trait on to future generations.

Our own species has evolved in similar fashion under the influence of the Red Queen, but today we have also begun to rival her by driving the evolution of other species in unexpected ways.

During the 1970s, Zambian fisheries managers learned that their local fish communities could change rapidly through evolutionary cause and effect. When a national park was established in northern Zambia, it protected the crocodiles of Lakes Cheshi and Mweru Wantipa from hunting. More crocs survived into breeding age as a result, and the population boom that followed caused more intense competition for fish. When their normal prey ran low the crocodiles began to raid gill nets offshore, shredding them in the process.

No longer able to shoot their croc competitors, local tilapia fishermen responded by moving closer to shore and using cheaper, fine-meshed mosquito netting, a new form of natural selection that further complicated the situation. Small fry hide from predators in the shallows, so the finer nets caught them as well as the adults. Fewer tilapia reached adulthood as a result, and those who did so matured more quickly and were therefore smaller, on average. As a result, puny fish with "live fast, die young" genes became the norm.

Shrinkage under fishing pressure is now increasingly recognized as a fundamental evolutionary process in fisheries all over the world, and genetically based behaviors are changing, too. Biologists in Connecticut recently noticed that fewer largemouth bass were being caught in certain lakes even though their overall numbers remained high. Closer study revealed a surprising reason for the change. Individual fish with faster metabolisms were more energetic and therefore more likely to strike a lure and be removed from the population. The metabolisms of bass from heavily fished lakes had therefore become slower on average than those of fish living in protected waters. With more than three hundred thousand anglers working the state's waters in 2011 alone, this kind of selection favored lazier, more sluggish bass—survival of the fattest, if you will.

A similar study by German scientists found that selective angling can fill a lake with fish who are not ideally suited to survival in the wild. Large, bold bass are generally the most attractive to mates, most likely to fend off an intruder, and most likely to raise lots of offspring because they defend them against nest predators. Anglers, however, can rewrite the rules of selection for bass by turning large size and boldness into liabilities and thereby favoring removal, rather than survival, of the fittest individuals. The researchers noted that artificial stocking can keep the total numbers of bass in the lake stable, but it can also hide declining overall health among the fish themselves. The result is a "Darwinian debt" that evolution can theoretically reverse, but only over many generations.

I recently became aware of my own potential as an agent of evolution among largemouth bass while canoeing with my wife in the Adirondacks on a calm, sunny day in late spring. Several feet beneath our boat, pale circles dotted the bottom where male bass had scooped nests into the loose, pebbly sediment with their snouts and mouths. The fish who hung motionless above the nests were large, and my first thought upon seeing them was, "Where's my fishing rod?" But as we glided silently over them I noticed something else.

Each male bass was actively guarding a territory as he hovered in place, fanning his pectoral fins and thereby aerating thousands of eggs that had recently been laid by his mate. However, different individuals responded differently to our approach. In most cases the guardian fled before our shadow crossed his nest, but some did so more reluctantly than others. A few even stood their ground and glared up at us as if daring us to challenge them. Those bolder individuals who were willing to face down a canoe to protect their eggs might also have been the most likely to bite an intruding lure. If I had used a rod that day their boldness would have worked against them in the struggle for existence, and I would have missed a more rewarding experience.

What does all of this have to do with the origin of species? The incremental changes just described, which specialists call "microevolution," do not necessarily turn one species of fish into another. More often, traits vary slightly within a population around some average, or change in ways that do not alter their classification as species. Even my creationist friends who profess no belief in evolution acknowledge the role of natural selection in the spread of antibiotic-resistant bacteria and pesticide-resistant insects, although they prefer to call it "adaptation" rather than use the "e-word."

Given enough time and genetic change, however, an isolated population can become different enough for its members to stop interbreeding with the original population even if they are reunited. That transformation of one reproductively distinct kind of organism into another is "macroevolution," the origin of species.

THE PRIMARY FOCUS of our lake research in 1981 was not evolution but geological history. Lake Malawi sits atop a stack of mud, diatom frustules, pollen grains, and other detritus roughly 3 miles (4–5 km) thick that took 10 million years or more to accumulate. We would map and analyze some of those sediments, but as an avid fisherman I couldn't help noticing the cichlids in Nkhata Bay harbor while loading provisions onto the trawler that served as our floating lab. Hints of gold and indigo flashed up through aquamarine water where cichlids pecked at the algae-coated concrete of the pier, but angling for them would have been a waste of bait, as I later learned by wetting a few lines. Most of them were vegetarians, and finicky ones at that.

The struggle between predators and prey affects cichlids as it does any fish, but dietary fussiness also contributes to their success in Lake Malawi. So, too, do an instinctive sense of beauty and a primal desire for the opposite sex. One of the most interesting ways to see why that is so is to study cichlids underwater, as I remember doing several years later when biologist Peter Reinthal invited me to join him at Cape Maclear, a day's drive south of Nkhata Bay.

On a sunny afternoon in July 1988, I found an inviting beach scene beside a sheltered cove on the tip of the sparsely wooded cape. The aromas of wood smoke and ripe fruit greeted me as I parked beside a rustic hotel and bar where scruffy backpackers mingled with local fisherfolk. Most wildlife-lovers who visit Africa have lions and elephants in mind, but fascinating creatures also inhabit the lakes, and Cape Maclear was a fine place in which to see them. Submerged boulder-fields beside the eastern headland were part of the world's first freshwater national park and had been mapped on waterproof trail guides for snorkelers and scuba divers to follow.

The next morning, Peter and I rode an inflatable motorboat to the headland and tied up beside a jumble of boulders that dissolved into the blue-green depths. We slid our face masks into place, rolled backward over the side with a splash, and became weightless. My first breath underwater startled me as always but I quickly adjusted and followed Peter down to the bottom, trailing plumes of silver bubbles.

Hundreds of hand-sized cichlids swirled around us in wonderfully clear, comfortably warm water. I knew no formal names for most of them apart from their inclusive Malawian name, *mbuna*, so I was

Cape Maclear, Malawi. Left: A snorkeling site at the rocky headland. Right: Sandy beach and settlement. *(photos by Curt Stager)*

left with color patterns to describe them. Iridescent azure with vertical black bars. Powder-blue flanks and a canary-yellow belly. Black and white racing stripes from nose to tail on a buttercup background. As the mbuna fluttered beneath us like aquatic butterflies, ribbons of wave-bent sunlight rippled over boulders that were encased in fuzzy meadows of golden algae that glistened with tiny pearls of freshly made oxygen.

It seemed surprising that the rocks were not picked clean by the fish, but the algal gardens grew as quickly as they were cropped. Their thin veneer masked a vigorous productivity that the bubbles betrayed. Long ago when all life was microbial, the evolution of oxygen-generating photosynthesis doomed species who couldn't use or resist the reactive fumes that now sustain fish and fill scuba tanks. We are all descended from survivors of the world's first biologically induced air pollution crisis billions of years ago.

The seemingly random movements of the mbuna were fine-tuned by natural selection. Each species was an expert in finding a particular kind of food, and their mouths were highly specialized tools. Many of the cichlids before me combed filaments from the algal lawns with tiny, rasping teeth, but some concentrated on the tops of the boulders while others foraged along the sides. A yellow-tinged fish with thick lips suctioned debris from the rocks, vacuuming up everything from fish droppings to loose clots of algae. A darker one with longer, floppier lips wedged its face into crevices to slurp small creatures from their hiding places. Other cichlids specialized in zooplankton or crunchable snails, and one predatory species feigned death to lure curious minnows within reach.

Peter dip-netted a gorgeous blue algae-scraper, swam over to me, and gently opened the fish's mouth for me to peer inside. Minute teeth protruded from a bony plate on the roof of the throat near the gills, an extra set of customized tools that mbuna use to grip, crush, or sieve their favored foods. Because of their strictly defined diets, many kinds of cichlid could share the same habitat without wasting energy in fights over meals and without overexploiting their food base. Competition can be as much a driver of selection as predation, and the benefits of avoiding it allow many endemic species to coexist in these lakes.

Not all cichlids are so cooperative, however. The fins of many of the fish were strangely ragged, the work of fin-biters. Most of the mbuna at Cape Maclear were of similar size, but if all you need is a bite of fresh flesh once in a while, then why not simply snatch a mouthful instead of tackling the whole animal? Several species did just that, waiting in ambush for an unwary passerby and then taking a quick snip. Others raked mouthfuls of scales instead, and at least one species was said to pluck eyeballs.

If there was a way to find any imaginable food in the lake with minimal competition, then it was likely that at least one kind of Malawi cichlid would develop ways to use it. But with so many crafty food seekers on patrol, how could helpless hatchlings survive?

One fish shepherded a cluster of tiny fry who resembled translucent grains of rice. When I swam closer she opened her mouth and inhaled them. She was not a predator, however, but a parent, and her mouth was not a trap but a shelter. Most African cichlids mouth-brood their eggs and young even though it can mean going without adequate food for weeks at a time. Carrying a tempting mouthful without swallowing it means that each hungry fish is not only a self-interested individual but also a member of an evolving species for which successful parenting is as important as eating. This aspect of evolution is called "kin selec-

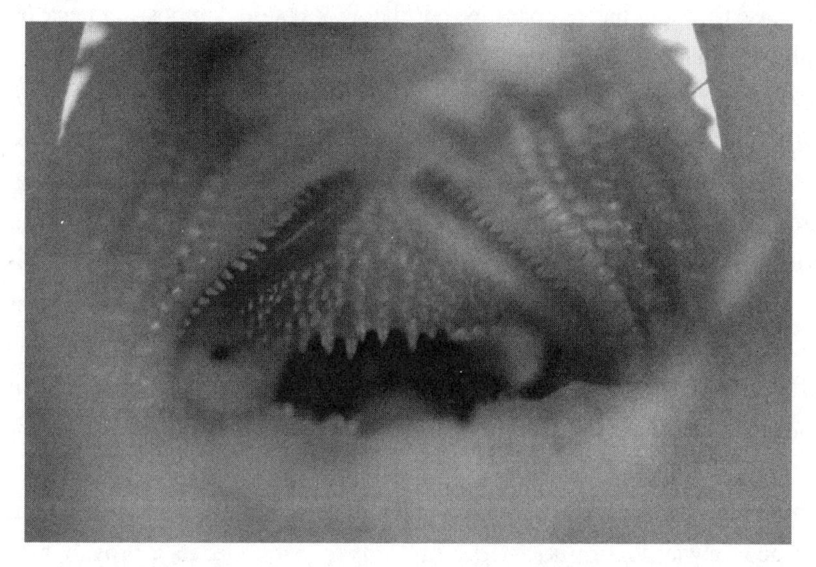

Specialized teeth on the roof of the throat of a Malawi cichlid above the toothy gill arches. *(photo by Curt Stager)*

tion," a form of natural selection that helps to hold young cichlid families together as well as our own.

It may seem strange that altruistic behaviors that we value so highly could arise through such mechanistic processes, but in a totally selfish world, life as we know it would be impossible. Kin selection is especially prevalent among social animals such as ourselves, and versions of it can operate among close associates as well as close relatives. Like other social primates, we rely on one another for survival and reproduction, and our ability to sense and anticipate the thoughts and feelings of others is as subject to selection as size and strength. We may call its manifestations empathy, shame, kindness, or generosity and imagine that they are uniquely human traits, but the works of Jane Goodall and other researchers in Africa have shown that our nearest relatives, the apes, share many of them, too.

One need not be a primate, however, or even very social in order for evolution to favor the care of one's own offspring. Crocodiles have thrived in Africa for millions of years not only because they are fierce but also because they are devoted parents. Mother crocs build composting nests on sandy shores in which decaying vegetation helps to keep their eggs snug and warm. When the babies hatch, Mom responds to their muffled cries by tenderly pawing them out of the nest, and she guards them against intruders until they are large enough to fend for themselves. Mother crocodiles with faulty genetic blueprints for parental behavior raise fewer babies, and selection tends to weed them out of the population.

The instinct to protect offspring is passed on to future generations as long as parents ensure the survival of their young, but the evolution of a defense in one species can also stimulate new ways of breaching it in another. One kind of Malawi cichlid with an upturned lower jaw secretly watches for fish whose cheeks bulge, a sign that they are mouth-brooding eggs or fry. The hunter then darkens or dims a broad lateral stripe in a way that mimics the target, making it easier to approach, ram the victim's face, and snap up whatever is knocked loose.

The feeding specializations that reduce competition among cichlids in concert with the survival boost from mouth-brooding allow many different kinds of fish to share a lake with minimal conflict. But in order for so many home-grown species to persist in Lake Malawi without melting back into a single communal gene pool, yet another evolutionary mechanism must operate as well, one that involves the art of finding a mate.

The creative tension between "sexual selection" and natural selection among cichlids generates strange and wonderful traits. It paints them in peacock colors even though camouflage would hide them more effectively from predators, and it produces outlandish behaviors that make sense only as an aid to reproduction. It also works on us as well as fish.

Consider a macho ladies' man who wears a gold chain and drives a red Ferrari to a dance club amid a fog of cologne. His appearance and behavior reek of sexual selection. If not for the female attention he wins, or hopes to, lover-boy's costly attire would be a useless drain on his bank account, his expensive car a mere magnet for speeding tickets, his dancing a waste of energy, and his odor a turnoff to his buddies at the bar. If some aspect of an animal's appearance seems ridiculous in terms of immediate survival, then sexual selection probably has something to do with it.

When Peter and I left the rocky headland and swam parallel to the beach, we crossed a sandy plain pocked with shallow craters the size of salad bowls, each a bower for love. Peter pointed to a bowl several body-lengths below us where a cichlid hovered with feathery fins waving. It was a territorial male showing off in an arena of his own making. Most of the passersby paid him no mind, but one rather plain-looking fish suddenly stopped to watch.

The female had recognized him as a potential mate by his shape and colors. However, each fish is a unique individual, and this fellow's moves and good looks in comparison to those of others of his kind had also tempted her to select him for closer scrutiny. Being a fish, she didn't analyze her attraction in terms of genes and evolution, but instead responded to a hard-wired appreciation for male beauty. Nonetheless, research by Dutch biologist Martine Maan shows that there may have been more to this fellow than met the eye. His exceptionally bright colors meant that he probably carried fewer parasites than duller competitors, and his courtship dance required vigor and coordination that only a well-made individual could muster. When natural selection and sexual selection work together in this manner, good genes in good-looking parents are more likely to become good genes in future generations.

If we had spied on the prospective couple a bit longer, we might have watched them circle one another in a seductive tango that was unique to their species. One false move and the thrill would be gone as surely as from a clumsy stomp on a dance partner's toes. If all went well, the female would eventually pause to deposit a fleshy orange egg in the nest. She would then scoop it up in her mouth to protect it before resuming her dance. She would lay another egg and then another until, caught up in the rhythm of stimulus and response, she noticed an egg-like dot on her partner's anal fin. She would then move in close as if intending to brood it. At that critical moment, a burst of semen from the male would fertilize the eggs in her mouth. Rather X-rated by human standards, perhaps, but that is how Malawi cichlids breed true in their crowded lake communities.

Such evolutionary marvels may seem to be the result of conscious change, but they are not. Cichlids do not develop fancy mouthparts *in order to* avoid competition with neighbors or seek mates *in order to* pass their genes on to future generations. They simply live out their lives, unaware of the larger story of their species, and evolution may or may not follow. It is we who, for the first time in history, give evolution conscious direction or purpose through domestic breeding and genetic engineering. Thus far, the evolution that we also induce through the management and exploitation of wild species has remained largely accidental, but that, too, is now changing as natural resource managers become more aware of it.

Most organisms carry enough genetic diversity and useful adaptations to handle normal changes in their environments, but when change is too rapid, severe, or novel it can push entire species into extinction. We are wise to remember this as our impacts on the world increase. Evolution is a mechanical process that cares nothing for the organisms it produces, and the history of life is full of mass death and destruction.

The sediment archives beneath Lake Victoria offer one such example of what severe environmental change can do to a lake and its inhabitants. They also show that we humans, after having evolved among the lakes of tropical Africa, have become an evolutionary force of nature in our own right, as well.

IF YOU COULD travel seventeen thousand years back in time, you could walk across what is now the bottom of Lake Victoria without wetting your feet. The last ice age was ending then, and tropical weather systems fell into disarray as the world warmed. The desiccation of Lake Victoria is a textbook case of natural selection gone rogue in which all of the lake's residents lost their races with the Red Queen. It also offers a dramatic example of how devastating climate change can be.

An early hint of that astonishing story came to light in 1960 when Joe Richardson and his fellow graduate student Bob Kendall collected sediment cores from Pilkington Bay, a 5-mile-wide (8 km) indentation on an island near the Nile outlet at Jinja, Uganda. Livingstone's boat had been lost in the croc attack at Lake Cheshi, so they strapped a sheet of plywood onto pontoons and used it as a platform from which to drive their sampling tubes into the lake bed. Their goal was to determine if ice ages had affected tropical Africa, and they would do so by analyzing microscopic fossils and minerals in their cores.

Ice ages in Africa? Admittedly, the idea may seem strange. A United States senator openly ridiculed Livingstone's lake research during the 1970s with a "Golden Fleece Award," a publicity stunt that was intended to expose supposedly wasteful spending by the National Science Foundation. In reality, his suspicion was ill-founded and the scientific value of the project was enormous. The ecological history of the Victoria basin, the subject of Kendall's doctoral dissertation, would challenge conventional thinking about global climate and evolution.

Tropical forests alone contain between one-half and three-quarters of all known terrestrial animal and plant species, and during the 1960s many scientists thought that such diversity could evolve only in stable environments. That hypothesis was undermined when the core sampler that Kendall and Richardson were using struck a hard layer 55 feet (17 m) below the mucky floor of Pilkington Bay. Dense gray clay beneath the softer mud in their cores meant that the surface of Lake Victoria had sunk low enough for the bottom of the bay to dry out. Later, Kendall also found grass pollen near the base of the longest core, dubbed "P-2," which showed that arid savannas once covered the landscape where lush forests now grow.

The P-2 record of drought, along with stranded high-water beaches around many of the African rift lakes, demonstrated that climate change does indeed affect the tropics even though continental ice sheets never touch them. Intense warming of land and sea at low latitudes is the heat engine of the Earth's climate system, and when ice ages cool the engine it tends to slow down. Less moisture evaporates from the oceans, and less warm, humid air rises and condenses over much of the tropics. Less rain falls there as a result. So much for the supposed stability of African environments.

Those were impressive results for a research program that won a misguided Golden Fleece Award. More surprises were yet to come after further

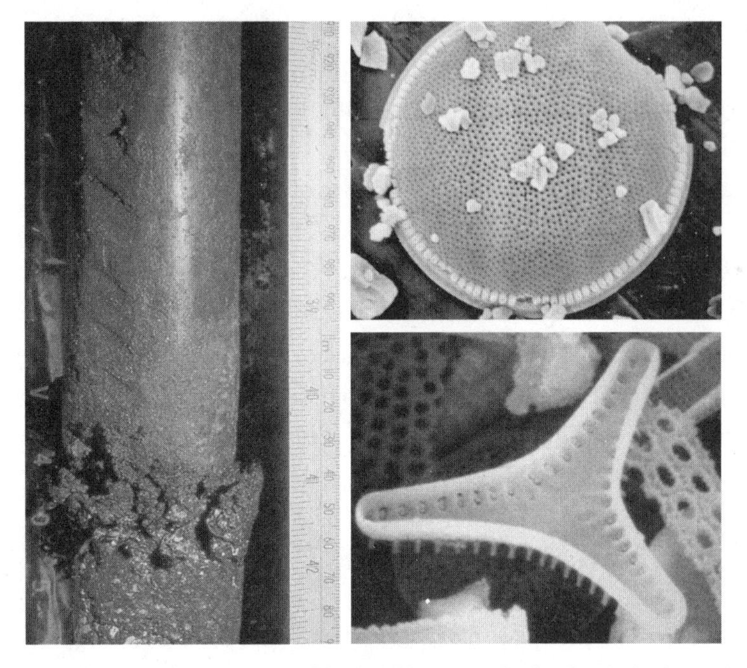

Evidence for the desiccation of Lake Victoria. Left: Sunbaked mud and beach shells at the base of a core. Right: Diatoms from the core that indicate shallowing (lower: *Fragilaria*) and salt buildups when the lake lost its outlet (upper: *Thalassiosira*). *(photos by Curt Stager)*

studies confirmed that the whole lake vanished seventeen thousand years ago in a killer drought that lasted for decades to centuries.

When Lake Victoria vanished, so did other lakes in tropical Africa and southern Asia during a near-collapse of the Afro-Asian monsoon system, a source of rainfall that now sustains billions of people. It was one of the most widespread and severe climate catastrophes in the history of anatomically modern humans. Genetic evidence of a contemporaneous population crash has recently been discovered among residents of India, which suggests that some of us may carry traces of it within our genes as African lakes do within their sediments. The precise causal connections have yet to be determined, but the great drought stands as a reminder that climate change can sometimes be rapid and devastating. It also warns us that similar risks could accompany our impacts on climate today.

For evolutionary biologists, the demise of Lake Victoria was especially noteworthy because of what it meant for today's cichlids. If the lake is relatively young, then its hundreds of endemic species must be young, too.

Relatives of the doomed Victoria cichlids probably survived in tributary rivers higher up in the mountains and in other lakes that were too deep to dry out completely. Their genes preserved a rich assortment of useful mutations from ancestors who had previously adapted to diverse environmental conditions. That genetic survival kit provided raw material for natural selection to act upon in a renewed lake from which large predators had been eliminated. In that ideal setting, the flock of new species diversified so rapidly that scientists have called their evolution "explosive."

To learn that Lake Victoria developed a new fish community after the dry period ended offers some hope in this modern age of human-driven habitat loss and extinctions. However, it would be wrong to shrug off today's environmental problems by saying, "change happened before, so what's the big deal?" The recovery at Lake Victoria came with two important caveats.

Refuges are crucial. Without wild places to serve as arks of genetic diversity, any evolutionary recovery that follows a major environmental disaster has fewer variants to work with at the outset and may even have to start from scratch.

Extinction is forever. The fish community of the rejuvenated lake was not simply a carbon copy of the former one, but an altogether novel one. In that sense, the complete recovery of a devastated ecosystem may be impossible.

With those thoughts in mind, I visited Pilkington Bay in June 2000, with four students of my own. Our aim was to investigate a dramatic new phase of the lake's evolutionary history, this time the result of human activity rather than ice ages.

When the captain eased the research vessel into the center of Pilkington Bay and dropped anchor, our gravity corer was swung over the rail on its tether, burbled as it filled with water, and sank beneath the surface. The bay was so full of greenish plankton that the sampler faded from view within 3 feet (1 m) of the surface. It was that clouding of the

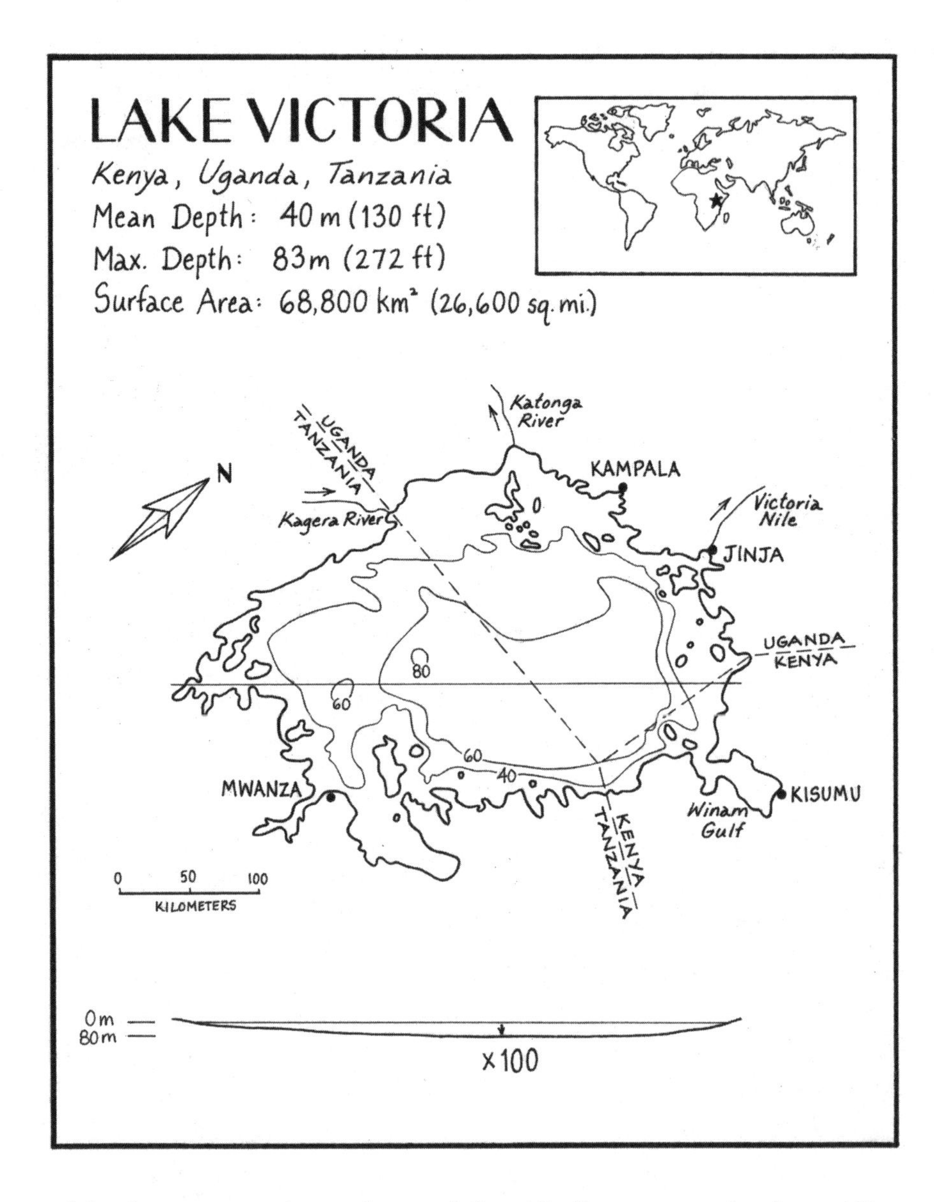

lake that we wanted to understand. Specifically, we wanted to know if it had happened because someone released Nile perch into Lake Victoria during the 1950s.

Nobody seems to know for sure who did it, but British colonial officials had long wished to enhance commercial fisheries in the lake, and *Lates niloticus* was their favored species. Whopper perch can reach 6 feet (2 m) in length and weigh nearly 450 pounds (200 kg), and the endemic cichlids, which many fisheries managers considered to be less desirable, would sustain them. The perch would be more exciting for visiting sport-anglers to catch, and their fillets would be more appealing to foreign diners than those of the bony little cichlids, bringing new sources of income to local entrepreneurs from global markets.

The idea was controversial, however, and ecologists warned that adding predatory perch to the food web could have unwelcome consequences. Losing small native fish who fed on phytoplankton might trigger top-down trophic changes and stimulate harmful algae blooms. Furthermore, the easily netted cichlids were a traditional source of affordable protein among local communities, and the perch would therefore compete directly with millions of impoverished people for the same cichlid prey.

The debate became moot when the first perch were caught along the Ugandan coast, and in 1962 and 1963 official stocking programs firmly established the newcomers. By the mid-1980s perch were everywhere, and cichlid populations crashed as predicted. As many as half of the five hundred known endemic species disappeared.

Fishermen who used to net cichlids from shore now had to use larger, more powerful motorboats in order to reach the open lake where most of the big perch were. Those who could afford the fancier boat, extra fuel, long lines, and bait necessary to capture the larger fish also had to hire crews to assist them, all of which helped a new cash-driven ecosystem to evolve.

Processing plants, smoking facilities, equipment suppliers, and make-shift fishing encampments sprang up along the shore, providing thousands of jobs. Perch fillets were flown on ice to Europe, Asia, and the Americas, pumping hundreds of millions of dollars into the local economy. For many fisheries managers the financial boom justified the perch introduction, but all was not well in the new Lake Victoria.

The lake became increasingly eutrophic as cyanobacteria and slender *Nitzschia* diatoms replaced other species in the phytoplankton. Microbial decay beneath the blooms consumed dissolved oxygen, forming dead zones on the bottom and killing fish. Was this the top-down trophic disaster that opponents of the perch introductions had warned of?

Our sediment cores suggested otherwise. The eutrophication had commenced at different times in different parts of the lake, but in each case it preceded the local perch boom by several years. A more likely culprit was phosphorus enrichment of the food web from the bottom up as more people became integral parts of the Lake Victoria ecosystem.

After independence came to Kenya, Uganda, and Tanzania during the 1960s, formerly landless people settled in the Victoria basin. Cities and villages swelled, and soil-trapping forests and wetlands were converted to firewood and farms. The perch boom drew even more people to the lake in search of employment, and by the end of the twentieth century some 30 million residents crowded the watershed, making it one of the most densely populated regions in Africa. Rivers and eroding gullies spewed phosphorus-laden sediment into the lake, turning immense swaths of it into chocolate milk. Even the air fed more algae-stimulating nutrients into the water until dust and soot from farms, domestic firewood usage, and fish-smoking facilities accounted for about half of Lake Victoria's annual phosphorus inputs.

By the 1990s, eutrophication, overfishing, and a shortage of cichlid prey had decimated the perch fishery. The drop in perch catches sent more people back to farming, so soil erosion worsened and washed even more nutrients into the lake. The perch collapse drove more people to net cichlids from the shore again, and in so doing they captured juvenile perch who sheltered there before they reached breeding age. That left fewer big perch to catch later on and fewer cichlids to feed them.

Newer regulations banned trawling, prohibited beach seining, and set larger minimum sizes for the fish that factories were allowed to process. As a result, perch are now more likely to reach reproductive age, but they continue to mature faster and at smaller sizes than before.

The Lake Victoria ecosystem is still evolving under the pressure of multiple human impacts, and so are the remaining cichlids. A copper-tinged species named *Yssichromis pyrrhocephalus* now carries thicker, heavier gills that draw more oxygen from stagnant, eutrophic waters. Recent forebears formerly fed on tiny zooplankton, but this new variety seeks larger insect prey in and around the oxygen-poor bottom sedi-

Severe gully erosion near Kisumu, Kenya, in 1988. Two people at left for scale. *(photo by Curt Stager)*

ments. Other species now have more streamlined heads and larger tails that may allow them to swim faster and escape the perch.

Where the water has become too turbid for fish to see clearly, accidental cross-breeding is melting formerly separate species back together. Swiss biologist Ole Seehausen has found evolution running in reverse among two closely related species of Victoria cichlid, one red and one blue. Because eutrophication has turned the lights out on sexual selection for them, the average male is now less colorful and the average female is less fussy, so indiscriminate mating now produces new hybrids who lack their forebears' original colors.

Only time will tell how Lake Victoria and its residents may change in the future, but it will largely depend upon how people treat the lake. Managing it like a protein mine is unsustainable, but ecologically sound stewardship of the huge watershed will require cooperation among nations and local communities as well as a comprehensive view of the entire ecosystem and humankind's place in it.

Like sparks of reflected starlight, the many species who live in and around the waters of tropical Africa are of different ages, having traveled different evolutionary routes to their present states over different time periods. Some are relatively young, some are ancient, but all form unique constellations of interconnected life. Our own species has long been part of those constellations, too, and it is increasingly so today among the ever-changing lakes of the Great Rift.

(photo by Kary Johnson)

5

GALILEE

Wildness reminds us what it means to be human, what we are connected to rather than what we are separate from.

—TERRY TEMPEST WILLIAMS's Testimony before the US Congress on the Utah Public Lands Management Act of 1995

THE BOOK OF GENESIS describes a lovely Garden of Eden where people lived in harmony with nature and where wilderness was a threat. In some newer versions of the story wilderness has instead become the paradise, and the domesticated modern world reflects humanity's fall. The differences between those opposing views of Eden underlie many of today's conflicts over environmental issues, but both share a dark view of the state of the world and, in particular, of humankind. Enter limnology. Sediment archives from lakes of the Holy Land preserve more history than the oldest scrolls and raise questions about some of the myths we live by. Did a place like Eden really exist? Is the concept of a lost earthly paradise relevant today? Are we better off without it?

In *The Power of Myth*, philosopher Joseph Campbell explained that myths are more sophisticated than fables and fairy tales. They help to illustrate our history and traditions, explain how the world works, and serve as guides to a meaningful life. By that definition, explanatory stories that science produces are not necessarily much different. A scientific approach to the lakes of the Jordan Valley might tell us, for example, whether the Sea of Galilee is really a sea or the Dead Sea is really dead, but a more fundamental "myth" of science underlies that approach, as well. It is that physical reality exists regardless of our ability to grasp it, and that a logical, empirical approach to knowledge is our clearest window on it. Astrophysicist Neil deGrasse Tyson recently invoked that foundational concept in a televised interview. "The good thing about science," he said, "is that it is true whether or not you believe in it." As science continues to upgrade our understanding of the world, our traditional myths can either evolve in response or run us aground on reality.

I visited the lakes of the Jordan Valley in 1988 as a scientist, but I was also fascinated by the interactions of fact and faith traditions that I encountered among the lakes there.

My personal introduction to the Sea of Galilee began with a visit to ecologist Moshe Gophen at the Kinneret Limnological Laboratory on the northwestern shore near Tabgha. It had been no small task getting into the fenced compound to meet him. My visit coincided with a violent intifada conflict between Palestinians and Israelis, and I had to leave my vehicle outside the gate as a precaution against car bombs. I was then interrogated by heavily armed soldiers who guarded a barrier of chain-link fence and barbed wire. The security concerns were themselves a sign that the so-called sea was not salty. As the largest nonartificial freshwater body in the region, some 13 miles (21 km) long and 140 feet (43 m) deep, it was not a sea in the usual sense but a lake, the primary source of water for the nation of Israel. A government-operated pumping station beside the lab was therefore a potential target for terrorists.

Gophen, then a middle-aged man with a neatly cropped thatch of white hair, met me at the lab. A light Mediterranean breeze whispered among palm fronds as he led me to the water's edge. White sails and motorboat wakes flecked the blue plain between us and the Roman-era resort town of Tiberias 5 miles (8 km) away. One golden, grass-carpeted

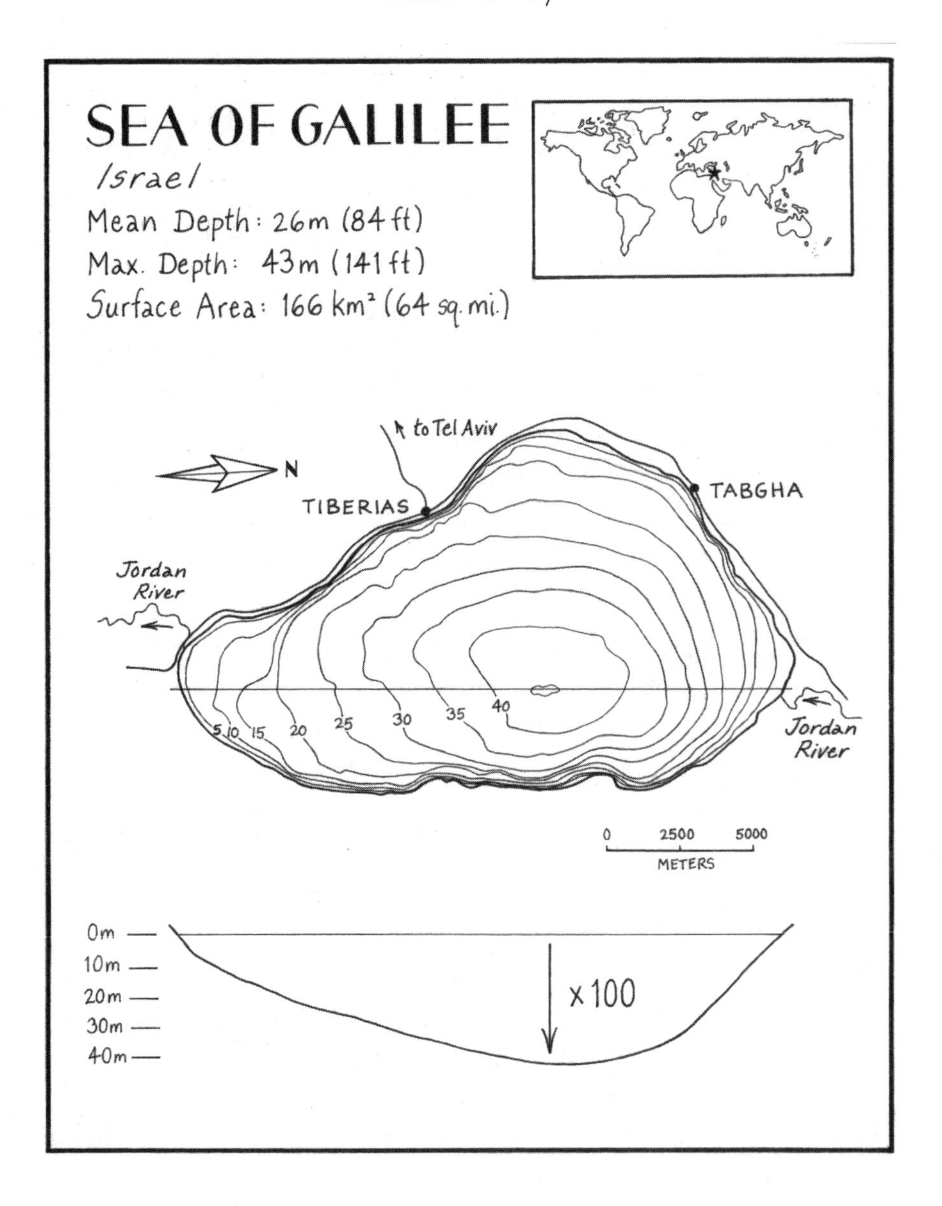

flank of the valley sloped down from the Galilee highlands to meet the shore there at roughly 700 feet (213 m) below sea level. Directly across from it, the opposite flank hung like a matching wrinkled curtain that was more faded by distance and summer haze. I knelt to touch the water and was mildly surprised at the bland taste when I raised my forefinger for a sample, even though I knew better.

Words can both reflect and shape reality, and simply calling this lake a "sea" can make people imagine it is salty when, in fact, it is not. It is hundreds of times less salty than real seawater, but the name fools people into thinking that it is a small ocean. For example, a best-selling scholarly book about Christ's ministry in Galilee describes the "salt air" along the "seacoast" at Capernaum, an unwitting flight of fancy by an otherwise meticulous author.

Confusion over the lake's label is a time-honored tradition. Early Greek versions of the gospels of Matthew and John gave it specifically marine names, including Sea of Galilee and Sea of Tiberias (*thalassēs*), but in the gospel of Luke it was the Lake of Gennesaret (*limnē*). The Hebrew term for sea (*yam*) was also applied to it in biblical chapters, and early Arab and Persian scholars called it both Sea of Minya (*bahr*) and Lake of Tabariyya (*buhairet*). As these names and their associated stories spread throughout Europe, local languages confused things further. The Old English word *sæ* refers to an expanse of water without specifying size or salinity, and similar lake terms also exist in Danish (*sø*), Swedish (*sjö*), and German (*see*).

Among limnologists, however, the distinction is clearer. A real sea exchanges briny surface water with an ocean. A lake can be fresh or salty, but it lacks the direct marine connection. Such is the case with the so-called Sea of Galilee, because although the lower Jordan River drains it, the river abandons its water on the floor of a desert valley far below sea level. Today, many Israeli citizens use the biblical Hebrew name *Yam Kinneret* or simply call it "the Kinneret," thereby sidestepping the issue of definitions. That label has been traced to a Bronze Age settlement on the southern shore whose name is thought to refer to the harp-like shape of the lake (Hebrew *kinnor*, or harp).

Fish the size of my hand hovered beneath the reflections in the shallows. "You might recognize them," Gophen said, "from your work in Africa." They were cichlids, evolutionary cousins of the species I had studied in the Great Rift lakes, and their home lake in the Jordan Valley was part of the same tectonic rift system that runs through East Africa. The fish before me were not endemic, but they were nonetheless uniquely connected to the lake because of their name. They were Saint Peter fish, prominently featured in the New Testament and the most commercially valuable species in the Kinneret. I grew up on stories of those fish from Bible study during my childhood, and although I was no longer religious it took a moment for my science brain to reclassify them as cichlids.

A *Sarotherodon galilaeus* resembles a nested stack of quarter moons with a tail. The largest ones are more than 1 foot (30 cm) long with spiny dorsal fins that their Arabic name, *musht* (comb), refers to. Both males and females brood eggs and young in their mouths, but neither sex is exclusively tied to a parental role. Some individuals share the work, some avoid it, and others monopolize it. The same kinds of fish are also known in Africa under other names, but they reach the northern limit of their range in the Kinneret, where they tend to congregate near warm springs that seep up through the bottom of the lake.

Gophen told me that many of the Saint Peter fish that were served in local restaurants were raised in farm ponds rather than the main lake, and in some cases the fish on the menu belonged to another species altogether. *Oreochromis aureus*, also known as blue tilapia, were similar-looking cichlids who were stocked in the Kinneret to supplement the annual catch. The two looked much alike on a platter, but Gophen claimed that the real thing tasted better. "Be sure to ask your waiter for *musht abyad*," he advised me, "if you want Saint Peter fish for dinner this evening."

Saint Peter fish take their name from a story in the gospel of Matthew. During his ministry in Capernaum, Jesus instructed his disciple Peter to pay a temple tax for him. "So that we may not cause offense," Jesus explained, "go to the lake and throw out your line. Take the first fish you catch; open its mouth and you will find a four-drachma coin. Take it and give it to them for my tax and yours."

The exact kind of fish is not specified in the Matthew account, but historians report that *S. galilaeus* was popular among diners in biblical times as well as today, so the presumed association sticks. Saint Peter fish are traditionally captured with nets, however, rather than hook and line. They are filter-feeders who sieve plankton from the water, and some scholars believe that the fish referred to in the Bible must have belonged to a predatory species that would be more likely to bite a baited hook.

Unfortunately, sacred status among fish and lakes does not necessar-

Saint Peter fish netted from the Kinneret. *(photo by Curt Stager)*

ily protect them from modern environmental problems. Back in the lab, Gophen explained that catches of *S. galilaeus* had declined recently due to a combination of overfishing, competition from stocked and invasive fishes, and nutrient pollution that favored cyanobacteria over more palatable algae. The problems began in earnest, he said, with the demise of a smaller lake nearby.

Lake Huleh (or Hula) was once the northernmost body of water on the valley floor, a shallow, marsh-rimmed expanse roughly 3 miles (5 km) wide that often flooded when the upper Jordan River swelled on its way to the Kinneret during winter rainstorms. For thousands of years it was a haven for cranes, pelicans, and other exotic birds who followed the valley on their seasonal migrations between Africa and Eurasia. It was also home to a rare cichlid (*Tristramella simonis*) and an endemic sardine-like fish (*Acanthobrama hulensis*). A Bedouin group known as Ghawarna fished, hunted, and farmed among the Huleh marshes, raising water buffalo and building their boats and dwellings from papyrus reeds.

During the 1940s, the Ghawarna were caught up in conflicts between

Arabs and Jewish settlers in the lowlands. In 1948, shortly before the British administration of Palestine relinquished power to the new state of Israel, a paramilitary operation swept several thousand Bedouin, including many of the Ghawarna, from the flatlands between Huleh and the Kinneret. Homes were demolished and many of the Bedouin fled north to Syria.

Most of the new settlers apparently saw the Huleh wetlands only as wasted real estate and a breeding ground for mosquitoes. Ditches and canals soon drained the water off into the Jordan River to make room for farms. Huleh's fish community was wiped out as a result, but the main focus at the time was on land, not wildlife. What at first seemed to be a triumphant conquest of the wilderness, however, soon became an environmental nightmare when the heroic myth met harsh reality.

The newly exposed lake bed was not fertile mud but fibrous peat that fluffed up and blew loose in annoying dust storms. The peat was also flammable and prone to smoky ground fires. Crops didn't grow well on the acidic duff, and when the Jordan flooded there was no longer a wetland buffer to stop the excess water from rushing into the Kinneret and dumping tons of detritus into the region's most important water source. In Gophen's opinion, it was that massive new source of nutrients that was most responsible for the onset of recent eutrophication problems in the Kinneret.

Pollution from runoff, sewage, fertilizer, pesticides, and herbicides poses less serious problems for the lake today and was already coming under control by the time of my visit. Stocking with hatchery-raised cichlids also helped to sustain the fishery. Gophen shook his head, though, when he told me that some fisheries managers wanted to release Nile perch into the Kinneret despite the hard-earned lesson of Lake Victoria. They expected the perch to grow fat on the native sardines who thrived offshore and to create a valuable new fishery. "Over my dead body," he muttered. "The food web here is already full."

Despite the many human impacts on the Kinneret, seasonal fluctuations in climate still control its annual cycle of stratification and mixing. In summer, a layer of warm water about 50 feet (15–17 m) thick floats atop the cool hypolimnion. Leakage from salt springs in the lake bed also makes the hypolimnion even denser and further stabilizes the layering. The deepest part of the lake can then become inhospitable to fish because its isolation from the surface allows the oxygen supply to run low and microbial hydrogen sulfide to build up and foul it. Midwinter chills the surface enough to allow the lake to mix and refresh itself temporarily before hot weather stratifies it again.

The strengthening afternoon breeze pushed its way into the lab through an open door as we spoke, and we stepped outside again to watch it sweep whitecaps across the surface of the Kinneret. Gophen described what was happening beneath the waves. "The wind is piling water up on the far side of the lake," he said. He held a sheet of paper horizontally to represent the thermocline boundary between the top and bottom layers and swept his hand over it. "When the wind blows hard and long from the west the thermocline tilts like this," he explained, dipping the downwind side toward the ground. The wind-heaped water squashed the hypolimnion on the far side of the lake and lifted it on our side, priming it for the seesaw motion of an internal seiche when the wind later died down.

Internal seiches within the Kinneret can rise and fall by 30 feet or so (9 m), much higher than the surface waves that conceal them. Their hidden sloshing stirs sediment-borne phosphorus and other elements up from the shallows and helps to feed the lake's algae, a nutritional asset to the ecosystem. Sometimes, however, they cause trouble for Saint Peter fish.

For some scientists, fish kills caused by seiche-driven upwelling in the Kinneret may explain a biblical miracle. In the gospel of Mark, Jesus wished to feed several thousand people who had gathered near the lakeshore to hear him speak. Attendees wondered how their limited provisions, seven loaves of bread and a few small fish, could feed them all. Jesus gave the meager rations to his disciples to pass along to the crowd, and the people miraculously "ate and were satisfied."

According to Gophen, a tilted thermocline might have contributed to that miracle of the loaves and fishes. "If a northwesterly wind blows hard all afternoon," he explained, "oxygen-poor water rises to the level of the springs near Tabgha where the cichlids gather. They begin to suffocate and the hydrogen sulfide in the water anaesthetizes them, too. Then all you have to do is wade out and gather as many floating fish as you want."

A clever idea, perhaps, but the gospel account also includes a multiplication of bread. How could seven loaves satisfy thousands of famished people? Gophen grinned. "Well, that would have to have been a real miracle, wouldn't it?"

In the years since my visit to Gophen's lab I have often heard people try to explain biblical stories from a scientific perspective. Some attempts work better than others. I've heard, for example, that Jesus walked on water to the astonishment of his disciples because a fluke of the weather allowed ice to form briefly on the lake. Such explanations can be entertaining as intellectual exercises, but they can also do a disservice to science and faith traditions alike.

The ice-walking hypothesis overlooks obvious practical details that, to me, make it no more scientifically convincing than the original story. Midwinter temperatures rarely fall below 40°F (4.4°C) at the Kinneret today, so winters two thousand years ago would have to have been much colder in order to form lake ice thick enough to support a person's weight, a claim that is not definitively supported by paleoclimatic evidence. It would also require that Jesus exploited a rare freeze-up at just the right place and time, that a visiting carpenter's son knew more about such a quirky phenomenon than experienced fishermen who marveled at the sight, and that he was able to walk sure-footedly on a slick ice floe while it bobbed on a storm-tossed lake at night, all without dying of hypothermia in the cold.

The great religious myths of this region are not scientific reports but the products of spiritual traditions that have inspired people of many cultures for centuries. At its core, the water-walking myth is not so much about weather and lake anomalies as about trust in hidden truths that seem to challenge common sense. One could say much the same about many of the findings of good science.

Mystical beliefs about lakes are common worldwide, and just as the surface of a lake reflects real features of its surroundings, some of those perceptions can reflect aspects of reality. Traditions associated with crater lakes that I encountered in Cameroon are typical. Monsters or spirits lurk in the depths and pull people under: a cultural warning of danger that also makes sense where crocodiles and accidental drownings are common. Sometimes the lake vanishes temporarily: a reasonable description of times when smooth reflections of the sky make the surface invisible. The lake has hidden connections to other lakes, the ocean, or a well-known mountain: not difficult to believe in the absence of depth contour maps, and technically correct as an interpretation of the hydrological cycle.

Some traditional knowledge is more accurate than the scientism of certain Western scholars. In 1986, for example, Cameroonian legends of exploding lakes gained respect among anthropologists when crater Lake Nyos killed 1,700 people with an explosive burst of magmatic carbon dioxide. Other traditional beliefs are demonstrably wrong, as was a common assumption among Barombi villagers that spirits, not fishing intensity or environmental conditions, determined the size of their catches. Nonetheless, even such factually inaccurate traditions might still have some cultural value if they can be adapted to reality when necessary rather than the other way around.

The complex interplay between mythic and scientific perspectives also applies to the lower Jordan River that exits the Kinneret and dies in the desert, as I later learned by following it to the Dead Sea.

THE JORDAN RIVER has been revered for centuries as the site of Christ's baptism. Many people who have never seen it imagine that it is deep, wide, and cold, and that it is therefore a refreshing place in which to bathe. In contrast, much of the lower Jordan was only an easy stone's throw wide during my visit in 1988, a shallow, tepid remnant of its former self due to the diversion of lake and river water for irrigation and human consumption. Much of it has become thick with marsh vegetation, green with algae, and tainted with pathogenic bacteria. So severe is the pollution from fields, fish ponds, and sewage in some lower reaches of the river south of the Kinneret that, in 2010, Friends of the Earth Middle East urged the Israeli government to prohibit tourists from being baptized in it because of the health risk.

Baptism in the Jordan River. *(photo by Curt Stager)*

The river is also a boundary between the nations of Israel and Jordan, and its meanders make spaghetti of the border. Dry washes dissect the parched land around the lower Jordan, forming networks of dusty gulches, or *wadis*. Long before the dawn of biblical history, a much larger version of the Kinneret covered this section of the valley all the way south to the Dead Sea, leaving laminated deposits behind when drier climates shrank it about seventeen thousand years ago. At its greatest extent, the swollen paleo-lake left bathtub rings of sediment on the steep valley walls hundreds of feet above today's surface of the Dead Sea. Geologists call it Lake Lisan, and its stranded sediments contain tens of thousands of years of environmental history. In more recent millennia, they also preserved some of the founding documents of Judaism, Christianity, and Islam.

In late 1946 or early 1947, Bedouin shepherds discovered the Dead Sea scrolls in a cave that was scooped into soft Lisan deposits overlooking a wadi at Qumran. One of the men is said to have tossed a pebble into the cave and been startled by the sound of breaking pottery. Within the cave were clay vessels containing documents written in Hebrew, including portions of Genesis. Archaeologists later found more texts in other caves that represented nearly every book of the Old Testament along with other religious and secular writings. Radiocarbon dating showed that most were written a little more than two thousand years ago, an age range that is consistent with bronze coins found among them. They are among the oldest known examples of Hebrew scriptures and are literary ancestors of the Bible and Qur'an.

Near the northwestern corner of the Dead Sea I parked beside a visitor center where the famed wadi gashed the dusty Lisan beds in Qumran National Park. When I stepped out of my air-conditioned car, it felt like walking into a furnace. Temperatures there routinely exceeded 100°F (38°C) in summer. Looking across the wadi at Cave 4 under the hammering Sun, I found it ironic that scrolls whose contents are respected in part because of their age were discovered within a geological archive that is much older but less often consulted.

The scrolls showed that many names have been applied to the Dead Sea throughout written history, some of which were as misleading as those of the Kinneret. Genesis mentions a "Salt Sea" in the Jordan Valley, and the "Dead Sea" label stems from *Maris Mortui*, a legacy of Roman times. Nonetheless, the Dead Sea is not a sea either, but one of the world's saltiest lakes, roughly ten times as saline as ocean water and just a few evaporative steps from slush. It is about 30 miles (50 km) long and 1,000 feet (300 m) deep, and it lies 1,410 feet (430 m) below sea level, lower than any other lake on Earth.

Continuing south from Qumran, I followed the winding shoreline road with high cliffs soaring above me on my right and a smooth sheet of water gleaming to my left. I pulled over amid shimmering heat waves and approached the lake on foot. A broad belt of turquoise in the shallows contrasted with darker depths offshore. It reminded me of white sand glowing beneath the waters of a coral lagoon, but the sand was made of salt, not coral. All around me cubic crystals of halite encased brittle weeds and dead twigs like rock candy, and a sparkling white crust crunched underfoot as I picked my way to the shore.

What makes the Dead Sea so salty? The answer lies on the atomic level. Groundwater and river runoff leach sodium, chloride, and other mineral ions from rocks and soils within a watershed of 16,000 square miles (41,440 km^2). When those solutes reach the Dead Sea, the water that carried them there evaporates and leaves them behind. A similar process left white salt stains on my thin cotton shirt as the thirsty air licked my sweat away during my walk to the shore.

The water was ice-clear. When I dabbed a bitter drop on my tongue, the taste convinced me that no normal plankton could survive in it, much less larger creatures. The Dead Sea's unusual bromine- and magnesium-rich chemistry also makes it particularly inhospitable to most aquatic life, and if one thinks only in terms of fish then the historic water body appears to deserve its name. In reality, however, the Dead Sea is far from dead.

What I couldn't see from the shore was a living world beyond the reach of my unaided senses. The water was actually full of microscopic prokaryotes, a group that includes true bacteria and their relatives, the Archaea. The drop I swallowed probably contained thousands of them.

Prokaryotes are the most numerous, diverse, and widespread organisms on Earth. They outnumber the stars in our galaxy if not the universe, and if they were all heaped into a single pile they would outweigh the human race. Like every other species from lilies to locusts, we are

Left: Salty crust on the shore of the Dead Sea. Right: Highway near Qumran, with the Dead Sea on the left. *(photos by Curt Stager)*

descended from early forms of these microbes. Their first appearance in primordial waters marked the dawn of life, and for two billion years they had the planet all to themselves. Today prokaryotes still live in every imaginable habitat from the dark, cold mud of deep ocean trenches to boiling hot springs. They even colonize the surfaces and innermost crannies of our bodies, which contain more of their cells than our own.

We may think of prokaryotes as simple and primitive, but scientists probe their internal workings for solutions to modern technological problems. Research on one group of prokaryotes, the Archaea, has yielded clues to the design of molecular tools that can function at temperatures hot enough to cook eggs. The revolutionary technique for amplifying small amounts of DNA into more useful quantities (PCR, or polymerase chain reaction) employs an enzyme that was found in prokaryotes from the boiling springs of Yellowstone National Park. That "Taq polymerase" enzyme, named for its *Thermus aquaticus* host, helped to win a Nobel prize for PCR developer Kary Mullis in 1993.

In 2010, scuba-diving scientists discovered a rich microbial realm on the floor of the Dead Sea. Thermal springs in the lake bed are rimmed with thick mats of Archaea despite salt concentrations that would suck the fluids out of most species. The divers wore full face masks to keep the hypersaline water out of their mouths and eyes, and extra-heavy weight belts held them down in the dense brine. The water is also full of salt-tolerant halobacteria who form a diverse plankton community and whose names evoke the harsh conditions they thrive in, including *Halorubrum sodomense* and *Halobaculum gomorrense*. The only way to consider this lake lifeless is to discount Earth's dominant organisms as living beings.

Dead Sea prokaryotes are masters at balancing salt and water within their cells, but they are not the only species to face that challenge. Our own cells struggle with it constantly, and fish survive in lakes and oceans only through a constant pumping of ions that consumes much of their energy.

The body fluids of Saint Peter fish, for example, are saltier than the Kinneret, so the fish risk swelling up because osmosis drives lake water into the permeable tissues of their mouths, guts, and gills. Kinneret fish must therefore urinate copiously and drink very little, as do most freshwater fish species.

Marine fish face the opposite problem. Their body fluids are less concentrated than seawater, so they live in constant danger of shriveling and must drink like proverbial fish. Many of them also pump excess salts out through their gills and excrete highly concentrated urine. Pity the Pacific salmon who hatch in dilute lakes of Alaska and British Columbia, spend months or years in the ocean, and then migrate back to their home lakes to spawn. Their journeys are osmotic ordeals on the cellular level as well as feats of strength and navigation.

Migratory salmon are physiologically suited to deal with the salinity differences between oceans and freshwater lakes, but a luckless fish who washes from the Jordan River into the Dead Sea is doomed. The hypersalinity of the Dead Sea overwhelms any physiological response, and such victims quickly die. Halobacteria avoid that fate by pumping salts into their cells, which helps to balance interior and exterior chemistry.

When I stepped into the Dead Sea at Ein Gedi amid a throng of other bathers, I didn't shrivel like a prune. My impermeable skin pro-

The Dead Sea near Qumran. *(photo by Curt Stager)*

tected me as long as I kept my head above the surface, which was easy to do because the water was so dense that I floated like a cork. So buoyant is the human body in Dead Sea brine that a man next to me reclined semi-upright in it to read a magazine.

People have come to Ein Gedi for centuries, partly for fun, partly because the water is said to cure ailments, and partly for beauty treatments. Many of the waders around me were covered in black muck that reeked of rotten eggs, an odiferous mess that forms amid geothermal springs where prokaryotes consume organic matter and release hydrogen sulfide. The fetid goo is supposedly good for the skin, and the Egyptian queen Cleopatra is said to have come to the springs of Ein Gedi during the first century BC to take the waters and slather the slime.

The Lisan beds show that the geological story of the Jordan Valley is much deeper than scriptural history. Other scientifically documented relics of that story also shed light on who we are as a species and on sacred lore of the Holy Land.

Biblical tradition describes a miraculous parting of the Red Sea, a truly marine water body south of the Dead Sea, but geology divides it lengthwise rather than opening a dry passage across it. In technical terms, it is a tectonic spreading zone in which magma boils up from the mantle and fills cracks in the broken sea floor. It widens by an average of nearly 1 inch (1–2 cm) per year because the tectonic plate of Saudi Arabia is drifting slowly northward like an iceberg on an ocean of molten rock. While the Red Sea widens, the western edge of the Arabian plate slides northward against another plate on the Mediterranean coast. Within that contact zone lies the Jordan Valley.

Geologists who study the valley see clear evidence of previous plate movements in its walls. The west side near the Kinneret is rich in limestone and flint deposits, but their counterparts on the opposite wall lie more than 60 miles (100 km) to the north. Where cracks cross the rift floor, the Arabian plate's motion drags them open, a process that caused the Kinneret and the Dead Sea to form. The Kinneret pull-apart basin holds several vertical miles (5–8 km) of lake mud, river deposits, and debris that tumbled in from hillsides during the last four million years or so, a suitably ancient stage for a long human history to play out on.

In 1959, an Israeli farmer found stone tools in the soil near Tel Ubeidiya, a grassy mound beside the Jordan River not far from the Kinneret. Ottoman settlers had built a small village on the mound during the nineteenth century, placing it atop the remains of a twelfth-century town that Crusaders had built on the remains of a biblical-age settlement. It was not the mound itself, however, that contained the oldest artifacts but sediments to the west of it.

Archaeologists later unearthed thousands of flaked implements, including ancient teardrop hand axes like those found in the East African rift. Among the tools were the fossilized bones of hippos, giraffes, and another mammal with African roots: *Homo erectus*. At the time, they were the earliest hominin remains found outside of Africa. The Ubeidiya bones demonstrate that the Jordan Valley was one of the first stops on our ancestral exodus out of Africa a million and a half years ago.

Other deposits near the former Huleh marshes were found to contain stone tools that were made three-quarters of a million years ago. Among those artifacts were charred bits of wood and scorched soil, some of the world's oldest evidence of the controlled use of fire. Tiny bones and teeth also documented the earliest known case of fish-eating and showed that Saint Peter cichlids were on the menu long before the time of Saint Peter himself.

The hunter-gatherers of Ubeidiya and Huleh were not the cave dwellers of our imaginations but people who mainly lived in the open, perhaps using ephemeral structures of brush and hides to shelter them at night or in harsh weather. That is not to say, however, that they didn't use caves at all. Evidence from caves in the adjacent highlands reveals how much like us our distant ancestors were.

Qafzeh rockshelter has yielded some of the oldest known evidence for a sense of life after death. Several people were interred in the shallow cave ninety thousand years ago, and their bodies were ceremonially decorated with seashells and red ochre. The remains of a child also indicated that the early residents of Qafzeh took care of one another in time of need. The little skull bore signs of damage and partial healing, which meant that someone must have helped the child to survive for several years after the injury.

Cultural diversity also has a long history in and around the Jordan Valley. Neanderthals occupied Amud Cave a few miles from the Kinneret fifty-five thousand years ago, when *Homo sapiens* lived in Manot Cave farther up on the shoulder of the rift. Their fossils alone do not necessarily confirm that the two groups interacted productively, but Neanderthal DNA in many of our genomes does. In addition to cooperation, however, our history of violence also runs deep. A fourteen-thousand-year-old human vertebra from Kebara Cave had a stone projectile point lodged in it.

The lake and cave deposits of the Jordan Rift Valley region show that basic human nature has not changed much over the ages. People of the past were much like people of today, clever and adaptable, often spiritually inclined, constructive in some ways and destructive in others. Our current problems cannot be blamed solely upon some sort of negative evolution in modern times or a recent blending of good and evil in human nature. But what about the land itself? How well does geological history fit with biblical accounts of it?

Most literal interpretations of the Old Testament put the age of the Earth within the 6,000–10,000-year range, which is very different from the 4.5-billion-year age that thousands of radiometric dates from rocks around the world have documented. The million-fold difference between the two timescales is akin to stretching an inch to 16 miles (25 km). Worldviews based on such a shallow puddle of written history are profoundly different from those that encompass the deep wells of geologic time. In a six-thousand-year-old world, for example, the origin of today's species by evolution is impossible not only because Genesis attributes it to a creator but also because there simply wouldn't have been enough time for it. In a 4,500,000,000-year-old world, there is sufficient time for sky-high mountains to be worn down by raindrops, rise again, and wear down again, as well as for tiny prokaryotes to give rise to figs, fish, and fishermen.

The oldest rocks of the Jordan Rift region include sandstones and conglomerates that are hundreds of millions of years old. More recent geological records of the Jordan Valley document a prolonged inun-

Lake Lisan sediment deposits at Qumran. Left: Laminations, possibly annual. Right: Cave 4, where some of the Dead Sea scrolls were discovered. *(photos by Curt Stager)*

dation by Lake Lisan near the end of the last ice age, but they offer no sign of Noah's global flood four thousand years ago or at any other time. The geological archives of the rift are inconsistent with myths of a young Earth or a biblical deluge, but what about Eden? Even many nonreligious people believe that a benign paradise once existed in which humans lived in balance and harmony with nature. Was there ever such a place in the region and, if so, what happened to it?

MANY INTERPRETATIONS of Genesis place Eden in the Fertile Crescent of Mesopotamia, but some locate it in the Jordan Valley. Biblical accounts also say that Adam and Eve were told to till the land, so their expulsion would presumably have occurred no more than about ten thousand years ago, when agriculture began in the region. The Dead Sea scrolls were written a little more than two thousand years ago, so a midpoint of six thousand years makes a reasonable time frame within which to postulate a physical analogue to the Garden of Eden in the Jordan Rift Valley.

One of the region's most detailed records of vegetation and climate comes from a small crater lake in the Golan highlands northeast of the

Kinneret. Some interpreters of the Old Testament believe it to be one of the "fountains of the deep" that produced Noah's flood. Druze residents call it Birkat Ram ("High Pool"), and their folk legends describe it as the weeping eye of a sheik's wife who mourned the loss of her husband. To geologists, however, it is a *maar*, a hole that was blasted up through bedrock by a subterranean steam explosion. Birkat Ram is shallow and only a few hundred yards across, but the maar beneath it is more than 200 feet (60 m) deep and full of layered fossiliferous sediment.

In 1999, a team of German scientists pushed core pipes into the pile. They found no evidence of fountains or global floods in the undisturbed sediments, but they did find pollen from semidomesticated plants that grew near the lake six thousand years ago. Wild wheat and barley had long been harvested by hunter-gatherers for their nutritious seeds, but mutations among some varieties had changed the way the seeds clung to the stems. In most wild cereal grasses, the brittle stems that hold the grains to the stalks snap easily in the wind and scatter the seeds widely. In the newly domesticated versions the stems were sturdier and early farmers could harvest them more easily without losing the kernels. People had already become agents of natural selection in that mostly wild setting, assisting the spread of easily harvested grains through small-scale plantings.

The pollen record of Birkat Ram offers little evidence of major human impacts on the vegetation in our theoretical Eden six thousand years ago. In those days, soils and climate still determined most of what grew where, despite a long history of subsistence agriculture and foraging in the region. Tabor oaks dominated the forests, but people also gathered wild pistachios, almonds, grapes, and olives in the woods. Local climates were moister than today, and a wealth of lakes and wetlands stretched from Arabia to West Africa. The water-blessed Jordan Valley was therefore not as unusual six thousand years ago as it is now with vast deserts nearby.

During that "green Sahara" period, people used bone hooks and barbed spears to capture fish, hippos, and crocodiles in places that are now dust bowls and dune fields. What was then the world's largest lake, "Mega-Chad," covered roughly 135,000 square miles (350,000 km^2) of north central Africa, more than four times the surface area of today's

Lake Superior. Today's remnant of it, Lake Chad, is only one-third as large, and the exposed, parched lake beds beside it have become the world's single largest source of airborne dust. The formerly aquatic deposits are so fine-grained that seasonal winds hoist them aloft by the ton and carry them for thousands of miles, spreading the powdered remains of the ancient lake far and wide. In 1832, Charles Darwin collected some of the "impalpably fine dust, which was found to have slightly injured the astronomical instruments" as it settled on the deck of the HMS *Beagle* off the northwestern coast of Africa. Later microscopic analysis showed that it was full of the glassy shells of lake-dwelling diatoms. Iron particles from the dust fertilize plankton in the Caribbean Sea and rain forests in the Amazon, and the smoke-like plume shades the tropical ocean beneath it enough to inhibit the birth of heat-fueled Atlantic hurricanes that would otherwise form there.

Archaeologists have found that early hunter-gatherers of the Jordan Valley repeatedly adjusted their lifestyles to follow the rhythm of climatic cycles in the millennia prior to and during the green Sahara period. When conditions became somewhat drier and woodlands shrank in the valley, people spent more time harvesting wild fruits and nuts in upland forests and cereals in the lowlands. They also hunted gazelles, wild cattle, and wild boars, often with the help of dogs. During wetter times they became more sedentary and territorial, favoring a narrower range of forest-based foods but a broader range of animal foods. When large animals became scarce around settlements, hunters focused on small prey such as hares, partridges, tortoises, and fish. Climate-driven transitions between flexible-mobile and rigid-sedentary cultural periods often happened quickly because oral traditions transmitted ecological knowledge across generations long before writing was invented.

All of this became more complicated, though, when both climatic conditions and human relationships to the land underwent a massive transition between six thousand and five thousand years ago. Crop cultivation had gradually tied people more closely to smaller home ranges and encouraged stronger senses of property and territoriality. Religions became more focused on deities who promised success in battle, healthy families, and abundant food and drink. Among them was Inanna, a

Sumerian goddess of love, war, and fertility, and Ninkasi, a goddess of beer. People built mud-brick homes, traded widely, grew wine grapes and grains, and raised sheep, goats, pigs, and cattle. They also began to use copper and other metals for axes, weapons, and works of art. With more powerful tools, they were able to clear more land.

Meanwhile, the end of the green Sahara period roughly five thousand years ago crowded more people into the remaining river valleys and lake margins, generating conflicts as well as innovations. Some historians speculate that clashes with proto-Semitic climate refugees from northern Africa eventually destroyed the cultures of our hypothetical Eden.

With this combination of climatic and cultural shifts, the once sparsely settled wilderness began to look more like an artificial garden. By four thousand years ago, cities such as Uruk and Jerusalem had developed, the first kingdoms formed, and new religions evolved. Within another millennium, Canaanite cultures of the Iron Age were being replaced by sky-god traditions that later morphed into monotheistic religions based on the Old Testament. Hebrew writings from that time refer both to care for the poor and slavery, manifestations of the familiar mixture of good and evil in human nature that also predated our hypothetical Eden.

Populations grew and people stayed put longer and in even larger numbers than before. Human impacts on the land and lakes intensified as olive orchards and vineyards replaced forests. Sediment cores from the Kinneret show that topsoil and nutrients from farms, livestock, and towns began to wash into the lake, making it more productive. That enrichment set the stage for further eutrophication when Huleh sediments and other modern effluents also began to wash into the lake.

In summary, the geological and lacustrine records of the Jordan Valley region suggest that a lush, somewhat Eden-like environment did exist in the Middle East about six thousand years ago, and that its subsequent demise was partly due to climate change and partly the result of human behavior. However, living conditions in that real version of Eden were more nuanced than in the benign paradise of legend.

The Jordan Valley of six thousand years ago was also an Eden for parasites and predators. People suffered and died from minor ailments such as infected cuts and diarrhea. Childbirth was often fatal for mothers, and most infants were unlikely to reach adulthood. The Huleh marshes bred malarious mosquitoes, and in ages past they bred crocodiles, too. In the biblical Eden lions lay down with lambs, but in the real world they ate them as well as human beings. The Jordan Valley of six thousand years ago might have been an interesting place to visit, but most of us would not really want to live there without the science, technology, and other benefits of modern civilization to support us.

Much of the Genesis version of Eden does not withstand rigorous scientific scrutiny, nor does the environmentalist myth of an idyllic wilderness paradise lost. The sharp dividing line that both narratives draw between humans and nature is misleading and also potentially harmful. In his essay *The Trouble with Wilderness*, ecologist William Cronon argued that "the romantic ideology of wilderness leaves precisely nowhere for human beings . . . to make their living." If people are considered a form of disease, then our planet can seem to be fatally infected and therefore no longer worth protecting, a recipe for unwarranted despair and inaction. Cronon suggests that when we pretend we are separate from nature in this manner, we "leave ourselves little hope of discovering what an ethical, sustainable, honorable human place in nature might actually look like."

The Jordan Valley's long history of compassion and conflict, cultural flexibility and rigidity, and variable impacts on the environment reminds us that finding our place in the world has always involved complicated choices with both positive and negative consequences. Real people of the past were as saintly and sinful—in other words, as human—as we are today, with a complexity that Walt Whitman recognized when he wrote, "I am as bad as the worst, but, thank God, I am as good as the best."

The core message in both versions of that history, however, is well supported by science: we share a common origin with all life and the Earth itself. Perhaps what we need today more than fruitless attempts to return to an imaginary Eden is for our ethics and worldviews to catch up to our newfound knowledge and powers as a force of nature. Then a real Eden might lie before us rather than behind us.

ON MY LAST DAY in Galilee I drove to the Mount of Beatitudes, a knoll that overlooks the northwestern corner of the Kinneret. The weather was unsettled, and gray whiskers of half-fallen rain dangled from scattered puffs of cumulus cloud. A sign near a chapel atop the hill told me not to wear shorts or walk on the grass. It was there, tradition has it, that Jesus gave his Sermon on the Mount.

Even though I had long since moved on from the Christian faith of my childhood, I found the scene more beautiful and moving because of its scriptural associations. "He leadeth me beside the still waters" in Psalm 23 surely refers to a lake rather than a river or ocean, and the author would have had only four candidate waters upon which to base the verse: Birkat Ram, Huleh, the Dead Sea, and the Kinneret. To me, the lovely harp-shaped lake before me was the obvious choice.

To my left was the river I once sang about in church camp. "River Jordan is deep and wide," the song went, "milk and honey on the other side." Not so deep and wide any longer, I had since learned, but still revered.

Below and to my right lay Tabgha, where the miracle of loaves and fishes is said to have occurred. A monastery there owned a spring-fed pool in which a tiny endemic species lived. Blind cave shrimp, *Typhlocaris galilea*, had evolved within dark limestone tunnels that fed the spring. Coincidentally, the similar Greek word *karis* also means grace or kindness, a fitting tribute to the shrimps' caretaker monks.

Out of sight over the Galilee hills stood Tel Megiddo, the site that the book of Revelations identifies as Armageddon, the final battleground at the end of the world. I thought of the many conflicts that have plagued this place through the ages. A million and a half years of human and prehuman presence in this, our ancestral gateway to the rest of the world, lends tragic irony to competing claims that "this land is ours alone."

Words from the Sermon on the Mount came to mind as I returned to

The Mount of Beatitudes overlooking the Kinneret. *(photo by Curt Stager)*

my car. "Blessed are the peacemakers, for they shall be called the children of God." Genes, fossils, and the principles of evolutionary biology are broadly consistent with such a view because they show that we are all relatives and made from the same earthly elements.

Our common humanity is rooted in a kinship that science now also extends beyond the boundaries of our own species. Our goose bumps still strain to raise the fur of early mammalian ancestors. Muscles at the bases of our ears still twitch reflexively at sudden sounds despite their inability to turn our ears toward them. And if we reach even deeper into the shared roots of all life they lead us to ancient prokaryotes much like those who live in the Dead Sea.

That unity of life extends all the way down to the atomic level. If we eat a wild-caught Saint Peter fish we consume the phosphorus of Jordan Rift rocks, the hydrogen and oxygen of cloud-borne lake water, and the carbon of a shared atmosphere. In doing so, we breathe life into the atoms of soil, water, and air as the fish did. Such insights are not just flights of mythic fancy but well-documented scientific truths that can resonate with some religious traditions, too.

Looking back on my visit to Israel as I write this nearly three decades later, I am pleased to know that the latest stewards of the Jordan Rift are working to turn the promised land into more of a land of promise.

Environmental organizations that formed in response to the draining of Lake Huleh have since helped to restore a small piece of it, now called Lake Agmon. It has become a major tourist destination and is once again a much-needed stopover for cranes, storks, and other migratory birds.

New desalination plants provide so much water for Israel that demand on the Kinneret has declined sharply since my visit in 1988. Municipal waste water is also being recycled for irrigation. There is now enough water left over in the lake's annual budget to keep the lower Jordan River flowing more reliably. Climate models suggest that these scientifically sound practices will become increasingly important as global warming boosts evaporation rates in the Middle East.

Plans are under way to revive the Dead Sea, which has been falling by several feet per year due to the shriveling of the Jordan River during the last half-century. Red Sea water is to be channeled downhill into the lake, driving hydroelectric turbines along the way to power desalination plants without producing excessive fossil fuel emissions. The "Red-To-Dead" project is not without its detractors. Some environmentalists worry that ocean water, being less salty, would deposit a surface layer atop the hypersaline lake and trigger unpleasant algae blooms. Nonetheless, the massive project represents a welcome example of cooperation between the Jordanian and Israeli governments, and the plans for it consider ecological consequences for the Dead Sea itself as well as political and financial interests.

Equally inspiring is a rising tide of concern for environmental issues among faith communities who have the language and authority to speak of care for creation as a moral imperative.

In 2015, Pope Francis released *Laudato Si*, an encyclical document in which he argued that it is not religious doctrine that causes our environmental problems, but rather our mistaken interpretations of it. Taking aim at readings of Genesis that lead people to treat the planet like a disposable commodity, he wrote that the exploitative interpretation of "dominion" over nature is a misreading of Scripture. It is our responsibility, he wrote, to "protect the earth for coming generations" and to recognize that we are part of creation and the global web of life. In that context, he argued, pollution, species extinctions, and artificial climate change are sins against God and our own human dignity. The pope has called nonreligious scientists "precious allies" in the struggle to care for creation, and that alliance is striking bright new sparks of faith and hope from the rock of science. "Nature cannot be regarded as something separate from ourselves," his encyclical states, "or as a mere setting in which we live. We are part of nature, included in it and thus in constant interaction with it."

The seeds of the future are in the words and myths we live by today. Whatever awaits us in that future, the lakes of the Earth will continue to reflect and record it as they have since the first people walked among the still waters of the Jordan Valley long ago.

(photo by Kary Johnson)

6

SKY WATER

The most dangerous worldview is the worldview of those who have not yet viewed the world.

—ALEXANDER VON HUMBOLDT (quoted by E. O. Wilson in *Half Earth: Our Planet's Fight for Life*)

SIBERIAN PINE RESIN spiced the crisp, clean air as our little ferry nudged a wooden dock on the shore of Lake Baikal, the world's deepest lake. I leaned over the rail and saw cobbles far below the gently rippling surface. Baikal is famed for its clarity as well as its size and depth, and visibility in the sunlit shallows often exceeds 80 feet (25 m). A creaking metal walkway was lowered, and passengers began to disembark onto the gravel beach. I waited while a dozen American high school students stepped ashore, bearing heavy backpacks. A dozen students from Moscow followed with three adults who would lead us into the wilds for three weeks.

It was August 1990, and what remained of the Soviet Union was opening to the West. My young charges, assistant leader Carrie White, and I were some of the first Americans to land on the beach at Khakusy hot springs, a rustic tourist retreat that was accessible only by boat. Having built my earliest impressions of Siberia around grim black-and-white TV imagery during the Cold War, I savored the clash between

Disembarking at Khakusy hot springs, Lake Baikal. *(photo by Curt Stager)*

my preconceptions and the reality that engulfed me. I was particularly struck by the strange familiarity of the place as we marched along the shore to our first campsite. The conifer-rich "taiga" forest was as green and fragrant as the woods back home in upstate New York, the locals were friendly, and the lake was beautiful.

I had come to Siberia as a co-organizer of the trip, but I was also on a pilgrimage of sorts. A visit to Baikal is on the bucket list of every self-respecting limnologist. It is a lake of widely reported superlatives, some true and some not.

Russians will tell you with pride that Baikal's deepest basin is approximately 5,387 feet (1,642 m) deep. True.

It is a trench-like rift lake nearly 400 miles (640 km) long, similar in shape and origin to Africa's Lake Tanganyika. True.

It is one of the oldest lakes in the world, born when its rift cradle formed at least 20 million years ago. True.

It contains 20 percent of the world's fresh water. Not exactly.

According to the United States Geological Survey, the approximately 5,500 cubic-mile (23,000 km^3) volume of Lake Baikal represents 25 percent of all the world's fresh water that currently resides in lakes. However, much more of Earth's fresh water is frozen, buried, or airborne. Ice sheets and mountain glaciers alone store the equivalent of 5.8 million cubic miles (24 million km^3) in solid form, and groundwater represents another 2.5 million cubic miles (10.4 million km^3). The atmosphere contains about 3,000 cubic miles (12,500 km^3) in vapor form, and if you could wring it all out of the air it would either drop an inch (2.5 cm) of precipitation over the entire planet or fill one-half of Lake Baikal. If you compare those numbers to the 321-million-cubic-mile (1.3-billion km^3) volume of the salty oceans, then Baikal's fraction of the global liquid budget is minuscule. Nonetheless, it is still a heck of a lot of lake, enough to submerge the state of Connecticut under a mile of water.

We hiked a short distance from the landing and set up camp within sight of the shore. A light drizzle began to fall while I helped the students to pitch their tents, and it reminded me how Baikal's gigantic mass of water got there. Simply put, it flew.

Lakes are transient eddies of distilled seawater that is air-mailed to them from the oceans little by little in countless droplets and flakes. Clouds and the seemingly empty gaps of humid air between them, rain and snow, lakes and rivers, berry juice and the moisture in our mouths, all are teeming with water molecules that evaporated from the oceans and will eventually return to them. Life on land is impossible without that cyclic flow, and it exists only where enough water falls from the sky to sustain living cells. A typical cotton-ball cumulus cloud contains about 500 tons (450 metric tons) of water, roughly as much as you could find within the bodies of ten thousand people. With water making up two-thirds of our body mass, we are literally elemental kin to clouds, lakes, and oceans that help to create the living world and hitch us to it.

More than 250 cubic miles (1,040 km^3) of water evaporate or transpire from plants into the atmosphere daily, and a similar quantity of fallen sky-water resides within organisms at any given time. The turnover of those molecules inside living beings is rapid, and in our case it happens to be close to the typical one- to two-week residence time of water vapor in the atmosphere. During the 1930s, the Hungarian scientist Gyorgy Hevesy used water labeled with deuterium isotopes to determine the

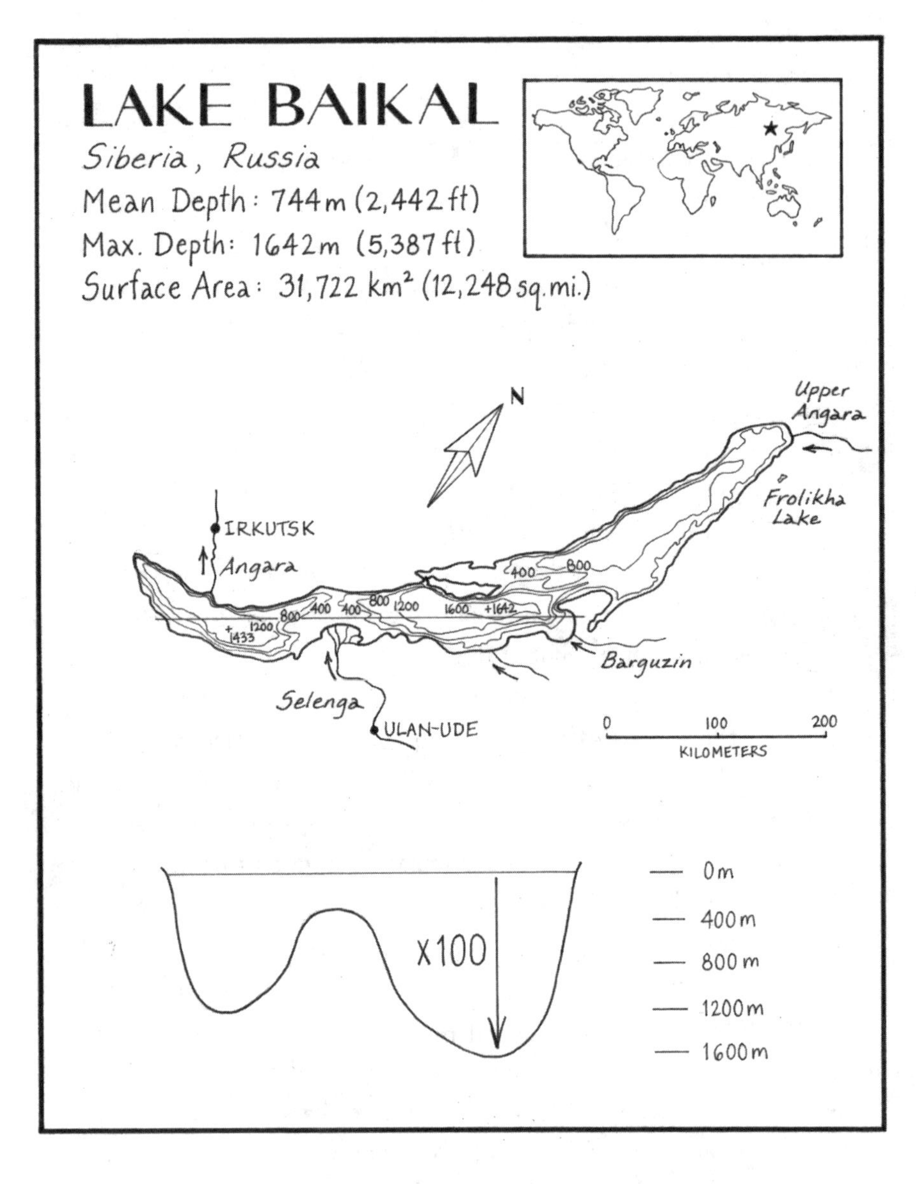

average residence time of a water molecule inside the body of a human being. Measuring the intermittent rivers that entered and left his own body, he found that the so-called "heavy water" could be found in his own urine about half an hour after he drank it. Applying the isotopic flow-through rate to the estimated billion-billion-billion water molecules within his tissues and organs, he calculated that any given molecule remained inside him for eleven to thirteen days.

The volumes of our bodies remain much the same from day to day despite the transience of our water, and the same is true of lakes. When limnologists measure the balance of inputs and outflows of lakes, they find a wide range of residence times that reflect depth, climate, and other factors. Some small lakes with relatively large tributaries and outlets flush within days to weeks, and others retain their water for much longer. Lake Geneva holds a water molecule for about a decade, on average. If that same molecule were later to land in Baikal it could be stuck there for more than three centuries. If it should then ride the lake's Yenisei River outflow to the Arctic, it might remain in the ocean for three thousand years or more.

Baikal hissed softly as the rain fed it, and the dimpled surface dissolved into gray fog where microscopic droplets floated like plankton in the chilled air, too small to fall. It would soon be dark, we were all soaking wet, and I needed to focus less on my place in the water cycle than on making a fire. It was important to reassure the American kids that they were in capable hands and to show our new Russian friends that they were not the only ones who knew their way around the woods.

I peeled strips of papery white bark from a nearby birch, knowing that they contained flammable oil that ignites even when wet. I'd done it a million times back home, and everyone gathered to watch as I stuffed the foolproof tinder under a heap of dead spruce twigs and set my lighter to it.

Nothing happened.

I tried and failed again several more times while the Americans looked increasingly forlorn. In contrast, the Russians were smiling, politely enough but in a way that suggested they thought I had lost my mind. Finally Gennadi, the real leader of the trip, squatted beside me and offered to help.

"This is not American birch," he whispered discreetly. "It is Siberian birch, and the bark doesn't burn so easily." When I asked how Russian campers made fire on a wet day, Gennadi nodded to Tim, the oldest student in his group. Tim reached into his back pocket, pulled out a scrap of half-melted Plexiglas, and handed it over. "This burns no matter what," Gennadi said. He set one corner of the scrap alight, held it under my heap of damp twigs until they caught fire, then blew it out like a gunslinger's pistol and handed it back to Tim.

My pride went up in smoke along with the crackling twigs, but I had also unintentionally demonstrated a central tenet of science as well as the main theme of the trip. Reality is not always what we think it is, and we can learn a lot from one another.

Early the next morning, the lake exhaled mist under a clearing sky. I heard stirrings in the tents so I couldn't linger on the beach for long, but I took a moment to savor the scene. The loose pebbles beneath my feet had been rounded by powerful waves, but for the time being the surface of Lake Baikal was so smooth that it resembled a fallen piece of sky which, in elemental terms, it actually was.

Beneath that mirror lived a rich community of creatures unique to Baikal. There were silvery *omul* fish down there, tasty and prolific enough to sustain a local fishery and the freshwater seals, or *nerpa*, who hunted them. Swimming among them were *golomyanka* fish, whose translucent bodies lack the internal air bladder that regulates flotation for most fish. Instead, buoyant oil comprises one-third of their body weight, enough that the fish are said to burn like candles when dried. The list of endemics also included more than 100 species of snail, 14 species of freshwater sponge, and at least 260 species of shrimp-like amphipod. Unlike the permanently stratified African rift lakes, Baikal supports prolific animal life at all depths because seasonal cooling and sinking of the surface mixes oxygen from top to bottom during the spring and autumn turnover periods.

Siberia has a well-deserved reputation for cold, and winter temperatures there often sink below -50°F (-45°C). Siberian summers, on the other hand, can be glorious, and despite the previous night's shower the day had been pleasantly warm and dry. An elderly fellow who sold me a loaf of bread at the landing joked, through Gennadi, that Siberian convicts beg local courts not to punish them with banishment to noisy, crowded Moscow. Baikal's climate also shielded it from vast continental ice sheets of the sort that bulldozed most of North America's temperate lake regions. Despite the deep freeze of Siberian winters, too little snow falls in that sector of Asia to build large, mobile ice masses. This helps to explain why Baikal's crop of endemic species is so rich: the crop has been growing and evolving for millions of years. In contrast, the recently glaciated Great Lakes of North America contain only a few unique fishes, mostly species of cisco.

Nothing struck my lures that morning, endemic or not, so fish were not on the breakfast menu. We gulped oatmeal instead, shouldered our packs, and headed inland on a gently rising footpath through the woods. Our destination was Frolikha Lake, which fed one of Baikal's tributaries. Gennadi had taken students there many times before, but this was his first joint expedition with Americans. The trip was the brainchild of my father, who had previously launched several other international youth exchange programs and camps, and Gennadi and I were learning to balance our shared duties as co-organizers while we got to know one another. He already knew the ropes here so I deferred to him in most things, but not all.

Cultural differences between the two groups were apparent at a glance. Most of our American camping gear was colorful, costly, and commercially produced. Most of theirs was drab and military in style, and much of it was homemade. We hiked in shorts, T-shirts, and lug-soled leather boots. They wore all-weather parkas and knee-high rubber waders. When the group paused to rest, drenched in sweat, the Russian kids glanced longingly at our water bottles. I urged my group to drink freely, but Gennadi warned his that drinking while hiking could induce dangerous cramps. The Russian girls also wore squares of foam padding strapped to their rears. We called them "butt-pads." When Carrie asked about them, Gennadi's colleague, Olga, said that the pads were there to protect feminine features from the cold ground while sitting. "Young women must think of their futures," she explained. American eyes rolled. We agreed to disagree and settled on separate rules.

We reached Frolikha several hours later and made camp. The comma-shaped lake, about 6 miles (10 km) long and 0.6 mile (1 km) wide, curled around the base of a ridge that separated two river valleys before emptying westward into Baikal. The main path ended there, but a narrow strip of boulder-studded sand between woods and water invited us to explore.

The rocks were horizontally striped with what appeared to be yellow paint. Gennadi said that it was pine pollen, which the forest produced

Frolikha Lake. *(photo by Curt Stager)*

in such abundance every spring that it turned the lake golden and caked the water's edge. The stacked bands represented sequential drops in lake level since the last pollen season. "It is very dry this year," he said, "because of El Niño." We didn't know it yet, but this particular El Niño event would become one of the longest on record, lasting several years rather than the usual few months. It would also bring unexpected blessings to us on our journey while wreaking havoc elsewhere.

I marveled at the link between this remote lake and a slice of warm ocean on the far side of the planet. Normally, Frolikha loses as much water to runoff and evaporation as the sky provides in a dynamic balance that keeps the lake's rim close to the forest edge. Every few years, however, tropical winds weaken and rearrange temperature patterns on the surface of the Pacific. El Niño floods strike the Peruvian desert, Indonesian rain forests dry out and burn, and the winds over central Asia keep water vapor aloft rather than dropping it, thereby starving Baikal and Frolikha of moisture. Invisible rivers of gaseous sky water are the lifeblood of lakes, and that year's flow to Siberia was weaker than usual or, at least, the vapor was not condensing into lake-filling clouds there at the moment. That gave me one less thing to worry about while camping with young folks in the back country. We wouldn't feel another drop of rain for the rest of the trip.

The next morning, we hoisted our packs again and stepped gingerly through a maze of rocks of various shapes and sizes, working our way slowly along the shore. We crossed the outlet river on a makeshift raft of logs and set up camp again. Later in the afternoon several of us began fishing from the beach. This launched an informal competition.

Gennadi's assistant leader, Sergei, landed a massive northern pike, or *shchuka*, so long that the tail almost brushed the ground when he hoisted his catch to waist height. I got nothing but a puny yellow perch. Back at camp, however, all fish were chopped into steaks and placed atop fresh alder wood chips in a sturdy steel box. After several minutes on a bed of hot coals, the box was opened and the aroma of smoked fish washed over us. Olga and the girls shredded the meat, mixed it with wild scallions, shaped it into small ovoid patties, and fried it. Delicious.

After dinner, I walked along the shore to watch the sunset color the water. A short distance from camp, I found another cluster of tents in the woods where several burly men huddled over a smoky fire. Their outfits were filthy, their faces were unshaven, and they outnumbered the adults in our party two to one. Unsure of the situation, I turned around in order to warn Gennadi that we were not alone. A hoarse shout rang out. "Amerikanski!"

I waved and continued on my way.

A louder "Amerikanski!" followed, then a rapid string of sounds akin to "DOBROPOZHALOVATZAVSTRECHU." I turned to face the strangers. They continued to shout and wave for me to come closer. I approached and explained that I didn't speak their language, but my English drew only blank stares and more Russian in reply. They beckoned me to sit beside their campfire, and I did.

Now the man who had first called out seemed even more determined to make me understand him. He tried various words and gestures, but

I could only shrug and smile. Suddenly, his grimy face brightened. He peeled a smooth sheet of birch bark from a fallen log, scrawled something on it with a charred stick, and handed it to me. It was the chemical formula for ethyl alcohol, C_2H_6O. I grinned and nodded my head. He beamed, and his companions cheered. Out came a tin of vodka, a can opener, and two plastic film canisters for shot glasses, and a new connection was forged between nations through the international language of science.

We humans misinterpret one another all too often, and the same is true of our interactions with lakes and the stories they represent. A global perspective informed by science can help us to interpret such stories more accurately, and digital imagery from above is an excellent source of those insights, as a brief virtual tour of the world's lakes will now show.

FOR ALL THE TROUBLE science sometimes brings in the form of weapons and ways to pollute the world, humanity is blind without it. Google Earth has accomplished wonders in that regard by opening our eyes to our home planet since its initial release in 2005. The free online program allows you to view a three-dimensional model of the Earth based on satellite, aerial, and ground-based photos that are regularly updated. You can examine and manipulate the digital globe from a distance or zoom in close to examine nearly any location of your choosing. With a click of a mouse it reveals most of the planet's surface pretty much as it is, regardless of what our imaginations or cultural traditions may tell us.

Consider, for example, the impact of the opening screen shot on some prescientific concepts of the world. It shows a sphere floating in space which the program allows you to turn or tilt in any orientation. Compare that to Old Testament cosmology that placed a flat Earth under a solid "firmament" dome of heaven upon which the Sun and Moon slid like mobile track lights. When technology such as this brings a myth into open conflict with science we must make an important choice. Are we willing to adjust the stories we live by when strong empirical evidence demands it, or do we cherish our illusions more than reality?

With Google Earth you can quickly and comfortably study lakes that only the most dedicated explorers could reach in days past. According to one satellite-based global tally you can potentially view 117 million lakes in all if you are so inclined. Twirl the virtual world on-screen and you can easily spot its three largest ones.

The huge Caspian Sea resembles the nearby nation of Turkey in size and shape. Most limnologists consider it to be the world's largest lake despite its name because it is fully landlocked. It is salty but not too much so to support commercial fisheries, and much of the world's caviar comes from Caspian Sea sturgeon. Despite the consensus among limnologists, however, the classification of this water body has yet to be determined in courts of law. If it is truly a sea in the marine sense, then its resources are theoretically open to all countries under the jurisdiction of the United Nations. If it is a lake instead, then Russia, Iran, and several other nations that surround it can legally divide it among themselves. And there is more at stake in the contentious debate than caviar alone. The sediments beneath the Caspian "whatever-you-call-it" are full of oil and natural gas.

Swing over to North America to see the largest of all freshwater bodies, Lake Superior, the westernmost of the Laurentian Great Lakes. They are themselves shrunken vestiges of former glacial lakes that could have drowned them all in meltwater at the end of the last ice age.

A great blue blob astride the equator in East Africa is Victoria, the world's largest tropical lake. Its irregular fractal edges follow the submerged contours of a broad plateau that has warped slightly downward and captured rivers that flow down from cloud-soaked highlands on the shoulders of the rift valley system. The islands in its northern sector are hills that project above the surface of the enormous puddle, a reminder that landscapes lie hidden beneath lakes as well as around them.

One can spend many hours exploring lakes in this manner and interpreting what their colors and shapes tell about them. Soda lakes in arid regions of Bolivia resemble pools of blood because local soils wash alkaline substances into them, thus allowing hardy red halobacteria to dominate the plankton. The famous turquoise tint of Lake Louise and similar lakes of the Canadian Rockies arises from the scattering of sunlight by powdery rock flour that melting glaciers pour into them. Sulfur particles turn New Zealand's volcanic Lake Ngakoro into mustard stew. Small crater lakes on the Indonesian island of Flores resemble green, turquoise, black, or brown eyes depending on the kinds of plankton or hot spring minerals that happen to color them at any given moment. Local residents attribute the changing colors to the shifting moods of ancestor spirits who dwell beneath the surface.

Some lakes are so devoid of plankton and dissolved substances that their water is breathtakingly transparent. In Crater Lake, Oregon, a white disc 3 feet (1 m) in diameter was still visible when lowered 144 feet (44 m) below the surface during a sampling session in 1972. As in the case of Baikal, its exceptional clarity stems in part from its large volume, which dilutes the few nutrients that wash in from relatively barren soils in a relatively small watershed and therefore supports little plankton. A lake in Tragoess, Austria, is also exceptionally clear because it fills with clean, cold snowmelt from the surrounding mountains for only a few weeks in spring, too brief a time for much plankton to develop. When the lake rises high enough to submerge a nearby park, scuba divers cruise like slow-motion birds above footpaths that lead them past empty park benches, green lawns, and healthy-looking trees in full foliage.

Telltale shapes often reveal how lakes formed. Lake Toba, Indonesia, was born in one of the greatest super-volcano eruptions in human history roughly seventy-five thousand years ago. The Toba blast covered southern Asia in 6 inches (15 cm) of volcanic ash and left behind a ring-shaped lake within a caldera bowl 50 miles (81 km) long and 1,657 feet (505 m) deep. A more ragged ring of water 39 miles (63 km) in diameter encircles a raised bruise where an asteroid smacked into eastern Canada 214 million years ago. The Manicouagan Reservoir of Quebec makes the impact crater more visible by flooding its sunken perimeter.

You may also notice patterns in the locations where lakes most often form. The first requirement for a lake to exist on land is, of course, water, and the atmosphere dispenses it in more or less predictable patterns around the planet. Mountains complicate the picture by squeezing moisture from the air on their upwind slopes, but in general rain and snow are most abundant where humid air currents rise and condense, particularly in a broad belt around the equator and in two smaller halos around the polar regions. The continents are therefore color-coded by the abundance or shortage of life-giving water in their respective climate belts.

Most of the equatorial regions, for instance, are leafy green when viewed from space. Flanking the central wet band are zones of sinking air painted in shades of red, brown, and gold that identify deserts and savannas where lakes are generally scarcer. Wind-generated climate belts have painted those three parallel color bands on Africa so the continent resembles a rainforest sandwich on desert toast, Sahara on top and Kalahari below. Ghostly traces of former lakes and rivers in the parched African deserts tell of shifting rain belts in the distant past.

Water alone is not enough to make a lake, however. Lakes need containers, too, and almost any kind of depression will do. The Great Rift Valley and rugged uplands of East Africa are full of them, but the huge, green Congo Basin in western Africa is too flat to retain much surface water. Rivers there send most of their runoff back to the ocean, and vast forests transpire much of the rest back to the atmosphere. Similar geography prevents many lakes from forming in the Amazon lowlands and instead sprinkles them among the peaks of the Andes.

Among the most lake-rich regions are those in the northern half of the northern hemisphere where retreating ice sheets have left irregular terrain. The air also drops plenty of water there, and local cool temperatures slow evaporation. Much of northern Eurasia, Canada, and the northern rim of the United States are therefore full of relatively young lakes, especially where glacial scour and blocks of dying ice left pits in deposits of sand and gravel. In some places, winds created lakes by pushing dunes around during droughty periods and then dropping rain and snow during wet periods until rising water tables filled the low spots between them. Hundreds of elliptical tundra lakes on Alaska's northern coast orient parallel or perpendicular to the prevailing winds, marking former dune fields and the sculpting action of air and water.

This basic lesson in reading lakes from above should make it easier to evaluate the features of a single water body in order to illustrate the tension between our imaginations and reality. Scotland's legendary Loch Ness will do nicely.

YOU KNOW THE STORY. An elusive beast lives in a lake, or *loch*, in Scotland. A famous photo of her long neck and head, purportedly taken in 1934, is burned into the imagination of everyone who has ever seen it, even though the picture was later shown to be a fake. She is widely presumed to be female, perhaps because her name, "Nessie," resembles Bess and Tess enough to sound feminine, although the *Ness* it stems from is simply the Gaelic name for a local river that feeds the lake. Most believers speak of *the* Loch Ness Monster rather than multiple monsters and envision her as a plesiosaur because of her sinuous neck, and because the lake seems deep and dark enough to harbor a holdover from the Mesozoic Era.

No specimens of a Nessie exist in museums, no photos have with-

Loch Ness. *(photo by Curt Stager)*

stood serious scrutiny, and no limnologist I know believes in her. Historians note that the first written account of a monster in Loch Ness was a legend about a sixth-century missionary to the Scottish Highlands named Saint Columba. He is said to have repelled a fierce water beast by making a sign of the cross at it, a bit of propaganda intended to impress the local pagans rather than skeptical scientists. Most other accounts are less than a century old.

For some of us, however, it is easier to believe in lake monsters than in the fallibility of our own minds and senses. In such cases, fact-based arguments against popular beliefs may actually strengthen faith in them. If seemingly heartless eggheads want to deny Nessie's existence and thereby deprive the world of her magic and mystery, then belief in her might even feel like a noble duty.

But what does the lake itself have to say about all of this?

If Google Earth takes you to Loch Ness, you will find it lodged in a rocky crease north of Glasgow. From above, the crease resembles a loose joint between landmasses, and that's just what it is. Four hundred million years ago, a rift much like that which Baikal now occupies split the British Isles away from North America and eventually widened to form the Atlantic Ocean. When it did, it sliced through a long east-west crack in the Earth's crust, stranding one loose end in New England near Walden Pond and the other end, the Great Glen Fault, in Scotland. Several other elongated lochs occupy the same crack, but most of us can't name them because they have no popular monster stories associated with them, nor do dozens of similar-looking lakes in the adjacent highlands.

Loch Ness is 23 miles (37 km) long, about 1 mile (1.7 km) wide, and 745 feet (227 m) deep, Scotland's second deepest lake. When I visited it in 2012 with Kary, we confirmed the mysterious appearance of the water by peering into it from shore. Unlike the more revealing water of Lake Baikal, it is dark brown and difficult to see through because it is stained with organic compounds that leach from peaty soils and wetlands in the watershed. Very large creatures could hide in those darkened depths, but Loch Ness is not unique in that. Monster-free Loch Morar, to the west, is about 250 feet (76 m) deeper.

Apart from presumed plesiosaurs, what other animals live in Loch

Ness? Copepods and ostracods (tiny shrimp-like animals with clamlike shells), brown trout and salmon and other fish, but none are particularly unusual or endemic to the lake. Baikal, in comparison, is full of endemics despite its location at a similar latitude on the same continent. The difference arises from the geological histories of the two lakes.

Unlike Baikal, the Scottish Highlands were repeatedly overrun during ice ages, and it shows from above. The smooth bedrock on the hilltops is a giveaway, as are the scoured valleys and pitted glacial deposits that allowed so many lakes to form there. The sediments under young Loch Ness tell a similar story with a relatively thin blanket of laminated mud atop coarser materials that the last ice sheet dropped before the lake formed. Ancient Baikal, in contrast, has more than 3 miles (5 km) of lake sediments beneath it that date back millions of years with no sign of major glacial disturbance.

To an unbiased observer, local geology and the nondescript species list indicate that Loch Ness is far too young to be a long-term haven for a seemingly immortal Nessie or a remnant population of plesiosaurs. If any such refuge exists, it would have to be located elsewhere. The many similar lakes nearby won't do for similar reasons. Plesiosaur fossils have been found all over the world in rocks that date back to the Mesozoic Era, including in Scotland, but none have been found anywhere in deposits younger than that. If we want to keep Nessie or other members of her species alive in secret for 65 million years, more than five thousand times longer than the age of any Scottish loch, then we must find an older, more stable habitat outside glaciated Scotland.

What about the ocean? Most plesiosaurs were marine, and some believers propose that one of them might have entered Loch Ness by swimming or flipper-walking along the Great Glen Fault from the North Atlantic just as marine seals occasionally do today. Baikal's nerpa seals could argue in favor of that hypothesis. They evolved from ringed seals who colonized the lake from the Arctic Ocean, so the premise is not necessarily far-fetched. The Caspian Sea also supports unique seals of similar origin, and Lake Saimaa, Finland, gained freshwater seals when the terrain between it and the ocean rebounded after the last ice age and trapped their marine ancestors inland. However, although seals are still common in the ocean and relict freshwater seals are found in several lakes around the world, Nessie's Mesozoic species is apparently restricted to a single individual or small population in a single Scottish loch.

With the combined testimonies of Loch Ness and Lake Baikal, we can now conclude several things. If Nessie is a plesiosaur, then she has been elusive for more than just the last century of false sightings and hoaxes. This would have to be a saga of deception involving millions of years of geological records the world over, a feat that would make her either an obvious delusion or more amazing than ever. To elude extinction when a planet full of other plesiosaurs died out, to survive as a breeding population through the ages without leaving a trace, and then to hide in a cold, dark Scottish lake bordered by well-traveled roads, homes, and businesses without being captured or frequently photographed while coming up for air, would be quite an accomplishment. The least we can do is offer Nessie a respectful refuge in our hearts if not in the loch.

This brief foray into the story of Loch Ness is incomplete by necessity, but it shows how much we can learn from lakes if we listen carefully to them, and also how persistent our beliefs can be. I confess that Kary and I are Nessie-lovers as well as Nessie-skeptics, and we keep a ceramic souvenir model of her on the window sill beside our dining room table. We both smile when we drive past signs that advertise "Champie," the knock-off Nessie of Lake Champlain near our home in upstate New York, and throughout my career I have often slipped a small offering into the water before collecting lake sediment samples. When possible, I have used whatever local peoples traditionally used to accommodate the particular tastes of their resident monsters and spirits. Among the Cameroon lakes they got whiskey. At Lake Victoria it was coffee. In Peru it was a tiny glass jar packed with Inca charms, and in Adirondack Mohawk territory the Underwater Panther gets corn or tobacco.

To me, acknowledging local traditions in this manner shows respect for people who have known the waters before me, even though I don't take the myths literally. It also makes me pause and collect my wits before engaging in potentially risky fieldwork, helps me to better under-

Face in the water. *(photo by Curt Stager)*

stand and appreciate my surroundings, and reminds me that there is more than one way to experience the world. In this manner, certain myths can serve and enhance our lives, even for scientists.

BEFORE MODERN TIMES the shapes, colors, and locations of most lakes reflected strictly mechanical processes. Soils stained them. Ice sheets buried them or choked them with silty meltwater. Volcanos and asteroids blasted their beds out of solid rock, droughts shrank them, and monsoons swelled them. Today we create, destroy, and alter lakes on similar geologic scales with at least some awareness of what we are doing, which adds an ethical dimension to those actions. It can be difficult to recognize the magnitude of our influence from the ground level, but studying lakes with the help of Google Earth can yield important insights into our place in nature and the condition of the planet.

Many lakes now show obvious signs of local human influence. Green swirls of floating duckweed on Venezuela's Lake Maracaibo. Cyanobacterial pea soup in China's Lake Tai. Algae blooms in the Gippsland lakes of Australia's southeastern coast. These and other discolored waters around the world tell of nutrient pollution from municipal sewage, livestock wastes, eroding soils, and fertilizer. Ironically, the crystal blue clarity of Lake Constance that makes it a source of safe drinking water for the German and Swiss residents of its coastline is also partly artificial. Phosphorus pollution from farms and municipal effluent that once clouded the water with algae has been so effectively curbed by environmental regulations that some anglers have begun to complain that the lake now contains too little plankton to support a large fishery. For them, Lake Constance seems to be a bit too clean.

Bathtub rings around lakes such as California's alternately shrunken or expanded Salton Sea and western Asia's shriveled Aral Sea betray varying intensities of rainfall and water extraction for settlements, farms, and factories. Erratic drowned shorelines and linear structures on the edges of water bodies, including the 3,300 square miles (8,500 km^2) of Ghana's Lake Volta, one of the world's largest artificial reservoirs, betray the presence of dams that have turned rivers into lakes.

During the last century, as many as 5 million new lakes were dammed or dug for water supplies, fish ponds, and other purposes in the United States alone. The creation of lakes by humans might seem like an opportunity for more aquatic species to thrive, but in some cases it can threaten biodiversity instead. A recent report in *Science* described the loss of unique river fishes to flooding of their habitats by hydropower dams on the Amazon, Congo, and Mekong Rivers. Perhaps surprisingly, losing certain artificial water bodies to infilling or neglect can also be a threat to species who have coevolved with them over centuries. In rural Japan, shrinking populations and changing diets are leaving untended rice paddy wetlands to fill in with sediment, thereby displacing rare dragonflies and other insects who once depended upon them.

Spinning the planet on a computer screen makes it easier to understand that everything is downwind of everything else and that air pollution affects every lake on Earth. The northeastern United States and much of Europe are afflicted by contaminated rain, though to a lesser extent now because of wise environmental legislation, because they lie downwind of major urban and industrial centers. A lake of milky, silt-choked meltwater has grown at the snout of the Tasman Glacier in New Zealand because fossil fuel emissions have heated the atmosphere enough to force glaciers into retreat worldwide. Swarms of egg-shaped lakes are creeping across vast sectors of tundra in Arctic Canada because global warming is softening the permafrost around their rims so that waves eat their shorelines more voraciously and push them downwind. Those changes fit a planetary pattern that has only recently been recognized with the aid of large-scale, long-term perspectives.

One of the most comprehensive studies of this kind thus far was published in 2015 by a large international team of investigators headed by Catherine O'Reilly of Illinois State University. The researchers compiled three decades of satellite-derived surface temperature measurements and direct sampling records to create a huge database of planetary change. Some 235 lakes were included in the study, representing more than half of the world's liquid fresh water. What they found was alarming. On average, lakes are warming faster in summer than the air above them and undergoing major ecological changes as a result.

The reasons for the faster pace of warming of lakes are somewhat unclear, but in certain cases it might be related to clearing skies. Some regions are becoming less cloudy, which lets the Sun heat surface waters more quickly. It might also be related to melting ice. Some of the most extreme changes are happening in northern North America and Eurasia where lakes normally freeze in winter. Warming has shortened the ice season there, which exposes lakes to the Sun for longer periods. Lakes that normally freeze have warmed by an average of 0.9° F (0.5° C) per decade since 1980, roughly twice as rapidly as ice-free lakes.

Some shallow lakes in Arctic Canada and Alaska are also drying out because they have lost the ice lids that once protected them from evaporation in summer. Others have become stratified for longer periods during ice-free seasons. Falling oxygen concentrations in their increasingly isolated bottom waters are releasing algae-stimulating nutrients that were stored in the sediments for thousands of years. The opposite problem afflicts large lakes in Africa where stronger stratification may now be trapping nutrients in deep waters beyond the reach of the plankton rather than allowing seasonal winds to stir them upward. At Lake Tanganyika, for example, a possible warming-related decline in algae at the base of the food web could threaten vital fisheries and livelihoods.

Another surprise from the O'Reilly study is how irregular the changes are. Lakes are not warming uniformly but in a piecemeal fashion that makes it difficult to recognize the overall trend without a global perspective. In other words, you can't necessarily tell what is happening by referring to your favorite lake in isolation, or even by comparing it to a few other lakes nearby. Depth, orientation with prevailing winds, and other features complicate the picture and can fool a casual observer into thinking that nothing is amiss. As O'Reilly said in an interview for *Science*, "ecological effects of global warming on lakes are not only imminent; in much of the world they are already under way."

Several days into our trek around Frolikha we were ready to push even deeper into the wild. Gennadi would lead us several miles up a tributary stream to an alpine lake high on the eastern rim of the watershed. All of us were excited by the idea, but I had also decided that I would never again let myself rely so heavily on a stranger's trip-planning methods. It was only on that morning that I realized how truly trackless the region was and how poorly prepared for it we Americans were.

Gennadi took me aside after breakfast to explain that the coming journey would be tougher going. The tattered map that he unfolded on his knee showed no formal trail and little flat ground for camping until we reached the lake. The map also lacked a name for our destination some two days' march from Frolikha, and Gennadi simply called it "Hotel Taiga" as he did our previous campsites. All well and good, but another feature of the map was more troubling. There was no trail indicated for our recent route around Frolikha, either.

"Where is the path you used," I asked, "when you brought your students here before?" Gennadi seemed puzzled, then realized what I was asking. They had simply followed the shoreline as we had been doing this time, he said, heavy packs and all. Without El Niño to lower the lake, however, the rubble-filled beach was normally shin-deep in water. "That is why my students wear rubber boots," he explained. "You were lucky to choose this summer to visit Frolikha with us."

Thanks to a fortuitous glitch in the water cycle, our Siberian trip would be more of an adventure than an ordeal, and I became one of the few people on Earth who was quietly blessing El Niño rather than cursing it.

The next morning, we followed a little-used path through the forest beside a clear stream where sleek, rust-colored fish darted like shadowy torpedoes. We paused to watch, and I wondered if they were young Siberian taimen, the world's largest members of the salmon family. Adults can reach 5 feet (1.5 m) in length and weigh more than 90 pounds (41 kg), so large that back-country guides advise anglers who pursue them to use whole mice for bait. In this region, taimen are traditionally thought to be the children of river spirits, and one legend tells of a party of men who survived the winter by slicing pieces of flesh from a taimen who had been frozen into the ice of a river. When the ice melted in spring, what remained of the taimen crawled up on land and ate the men.

Many of the mythic beliefs and rituals of indigenous Buryat peoples in the Baikal region reflect hard-won ecological knowledge that was developed through trial and error over thousands of years and transmitted orally from generation to generation. Some aspects of Buryat cosmology, for example, seem to reflect aspects of the water cycle that a hydrologist would easily recognize. Rivers like this one are said to be paths along which souls travel after death. When the souls finally reach the ocean they become birds and fly back to the river's source. Spirits of the most outstanding individuals also live in clouds and cause rain to fall or, as was the case this summer, to remain aloft.

Fish and spirits were not the only beings who shared the valley with us, however. When large clawed paw prints marked the wet sand near our last campsite, Gennadi reassured me that the local brown bears were

Hiking at Frolikha Lake. *(photo by Curt Stager)*

shy around humans. So, too, were weasel-like Barguzin sable whose dark, silky pelts were among the most valuable furs in the world. The trail we were now following might have been blazed by sable-trappers like the ones who left a windowless log hut beside the path. It was so small and claustrophobic that I wondered if spending long winters in it might have been harder on the men than the weather. Centuries of commercial trapping and hunting in these woods, both legal and illegal, had taught the wildlife to be wary of us, and apart from tracks we never encountered a nonhuman mammal larger than a chipmunk.

After a long day's march, the sinking of the Sun filled the valley with shadow, but we had not yet found a decent place to camp. Finally, we dropped our gear in a steeply sloping forest glade and tried to find a comfortable way to spend the night. The ground was rocky and uneven, difficult to pitch tents on and even more difficult to sleep on. Gennadi brightened our glum mood by giving this site a name other than the usual Hotel Taiga. "Hotel Govno," he growled with a wry smile. The Russian kids cracked up, and we Americans joined in when we heard the translation: the forest in the title had become feces, but we all felt a bit better about it now. We skipped the usual campfire, improvised a cold dinner, and let fatigue take us to bed early. In the morning, we knew, we would reach the literal high point of our journey.

We awakened at dawn and hit the trail, which grew increasingly steep. Several hours later we emerged from the wooded valley and crossed a wet alpine meadow to the edge of a lovely lake that I estimated to be about 200 yards (180 m) wide. According to Gennadi, we were the first Americans to visit it. We dropped our packs beside it, flopped beside them, and let chilly air wick the warm sweat away from our bodies. Olga lay flat on her back and held up what appeared to be a cross between a plump raspberry and a golden apricot, then popped it into her mouth. "Moroshka," she said, smacking her lips. Soon our metal drinking cups were full of the luscious fruit.

I was unable to find the mile-high lake on Gennadi's semi-useless map for future reference, and years later I still can't locate it even with the help of Google Earth, having forgotten its name. Clouds and snowmelt sustain many similar pools along the crest of the Barguzin Mountain Range, and our brief visit to this one would be an experience to savor on its own terms while it lasted as a choice moroshka of memory.

Alpine lake among the Barguzin Mountains. *(photo by Curt Stager)*

The little lake was not as spectacular as the view from a nearby ridge promised to be, however, and the kids soon headed over for a look. I sat by the water's edge a little longer, wondering what gift a Buryat shaman might offer to the ancestor spirits who are said to live in Siberian water bodies. Lake Baikal has spirits associated with it, too, including one named Lusud Khan, the Water Dragon-Master. Tradition holds that Lusud Khan once briefly became a woman and gave birth to the predecessors of the Buryat people. Today, some believers of the Loch Ness monster legend use the Lusud Khan story to corroborate their belief in Nessie, but I see more interesting parallels between Siberian folk culture and similar traditions elsewhere.

An ancient creation myth from this region tells of a sky spirit who descended to a primordial world covered with water. A loon plunged down to retrieve mud from the bottom so it could be piled up on another swimming animal to create land. Nearly identical creation accounts are common among many Native American cultures, including some in my home state of New York. In the Mohawk version, the mud was smeared on the back of a turtle. North America is therefore called "Turtle Island," and when we see ocean waves crashing against the shore it means that the great turtle is still swimming through primordial waters.

The occurrence of such similar creation myths on two continents is probably not a coincidence. Siberia lies on the western edge of what was once a vast expanse of tundra that bridged the Bering Strait during the last ice age. Stone age peoples hunted reindeer and mammoths there for thousands of years before retreating ice opened routes to the east through which early Native Americans colonized the continent, bringing their cosmologies with them. When Mohawk elders recite the story of Turtle Island they may echo the voices of Beringian ancestors as well as those of distant Siberian relatives today.

Ancestor spirits in lakes find surprising parallels in science, as well. The water molecules that form lakes also comprise our bodies, and the hydrological cycle distributes them worldwide. At least a few of the water molecules in any given water body are therefore likely to have flowed through the veins of some of our own predecessors and shape-shifted among their breath vapors, passing clouds, and the misty exhalations of lakes. Translate the English word for breath into the Latin *spiritus* and the presence of ancestral spirits in a lake becomes a scientific fact as well as a profound alliance of mind and heart.

Shouts of pleasure drew my attention to the ridge where the rest of the party was enjoying the view, and I followed. Wave upon wave of taiga-cloaked mountains stretched to the eastern horizon. Far below us lay a deep gorge through which a frothy white thread rushed to deliver water molecules to Lake Baikal. This was the farthest any of us had ever been from modern civilization. The nearest sizeable town was Ulan-Ude, about 250 miles (400 km) south of us. Far over the horizon to our left, taiga met tundra on the rim of the Arctic Ocean 1,300 miles (2,100 km) away. Directly ahead of us the dark sea of trees continued virtually unbroken for 1,000 miles (1,600 km) to the Pacific coast.

The moment was exhilarating, but also poignant. Even that immensity of trees and lakes was vulnerable now that the former Soviet Union was opening up to the world economy. Russian government officials and entrepreneurs were hoping to cash in on the global hunger for forest products by inviting other nations to clear-cut large swaths of Siberia's taiga, which is similar in extent to the great tropical forests of Brazil. Ancient Baikal was being tainted by debris from the rafting of logs to mills, chemical effluents from industry, and sewage from adjacent towns. That year's El Niño also demonstrated the sensitivity of local hydrology to climatic disruptions, including those of our own making.

Satellite data compiled by NASA have since confirmed that our concerns were well founded. The taiga that lay to the east of our observation point is now recognized as a deforestation "hotspot" where high-intensity logging, both legal and otherwise, and accidental wildfires are denuding thousands of square miles of woodland at rates comparable to those in the rapidly shrinking Amazon forests. A pulp and paper mill that used to dump tons of effluent into Baikal was closed in 2013 but, according to Environmental Wave, one of Russia's first environmental nongovernmental organizations, burgeoning tourist camps and boats are still fouling some of the more heavily developed sectors of Baikal's coastline.

During our visit to the little alpine lake, I also knew that our impressions of this scene were somewhat colored by romantic Western myths of untouched wilderness Edens. In fact, Buryat people had lived in those mountains for thousands of years, and some were apparently still living nearby, as a chest-high stack of stones suggested. In local tradition such cairns signify respect for spirit-inhabitants of special places, not unlike Thoreau's memorial rock pile at Walden Pond.

Beside the cairn I found the remains of a small campfire and two strange artifacts, a shiny metal bowl that fit neatly into my cupped hand and a flat strip of metal that could be used for scooping or mixing the contents of the bowl. In my mind's eye, a Buryat shaman sat beside the cairn. He mixed sacred herbs in the bowl with the metal strip and held the mixture over a smoky fire as an offering to his ancestors. I felt privileged to know enough about Siberian cultures to recognize the scene for what it surely was, a rare portal to a primeval connection to nature. Then Gennadi intruded.

"What are you doing?" he asked. I described my discovery, but he didn't seem to share my enthusiasm. I handed him the shamanic bowl and scooper, confident that a closer look would impress him. Instead, he smiled as he had when I tried to light the birch bark on our first evening in the taiga. Rather than stirring an imaginary potion, he tilted the strip slightly and pressed its narrow end against the outer rim of the bowl instead. My face reddened. It was a broken soup ladle left behind by a hiker.

Having just watched another mirage of my own making dissolve in the face of facts, I sighed and then smiled, too. This, I had learned, was the nature of the human condition. We are all vulnerable to self-deception. Interacting productively with others is one of the best ways to pierce such illusions, not in spite of our different perspectives but because of them.

This is also the way of science, a social enterprise that relies on group intelligence to correct errors by individuals. It is a way of listening to the world that requires humility and a willingness to trust solid evidence and logic more than the imagined voices and pictures in our heads. With the sacred soup ladle in mind, I brought that lesson and a useful myth home with me from the Baikal wilderness.

According to Siberian tradition, air gains psychic power within a person's chest and becomes a manifestation of the human spirit known as "wind horse." Strong wind horse helps a person to think clearly and see through deceptions, but using it for harmful purposes upsets the balance of the universe and ultimately depletes the wind horse itself.

Today, science has become a similar source of power that can serve us well if we respect and use it well. It is most powerful, however, when it not only informs the mind but is also harnessed to the emotional and spiritual wellsprings of our shared humanity.

(photo by Curt Stager)

7

HERITAGE LAKES

Hast thou named all the birds without a gun; loved the wood-rose and left it on its stalk?

—RALPH WALDO EMERSON, *Forbearance*

THE SUN SETS BEHIND a ragged wall of white pines, and both are reflected on the surface of Little Long Pond, deep in the Adirondack Mountain region of upstate New York. On this September evening I am sharing a canoe with Ted Mack, one of the most accomplished naturalists I know. Our wooden paddles lie on the floor of his battered aluminum craft and the air is as still as the water, but we glide slowly forward nonetheless. Ted has hooked something strong and heavy on his ultra-light fly line, and rather than risk snapping the delicate thread by reeling it in we are letting our own weight provide gentler resistance.

Ted makes his living as a librarian at Paul Smith's College as of this memorable moment in the early 1990s, but he has spent so much time outdoors among the woods and waters of the Adirondacks that he seems to know every bird, mammal, and fish personally. It was his idea to paddle and carry over three other lakes and footpaths in the Saint Regis Canoe Area because he knew that lake trout live here.

I am eager to see the mysterious creature who is dragging us, but I am also worried by the increasing darkness. It is a long way back to the car and we have no flashlight. Ted doesn't seem to mind, though, and his attention is laser-focused on the taut line in his hands. I am amazed that it hasn't broken yet.

Adirondack lake trout are chocolate brown with pale freckles on their flanks, and they are top predators in deep northern waters such as this. Beginning life as zooplankton eaters, they become fish hunters as they age and pursue nearly any prey smaller than themselves. They grow slowly over a life span of twenty years or more, but in lakes that supply them with sufficient food they can become both abundant and huge. One giant who was caught in nearby Follensby Pond weighed thirty-one pounds.

It is night now, but the rising Moon pours milky light onto the lake through shadowy pine boughs. From the front of the canoe, Ted tells me to nudge the bow into the wavering trail of reflections so he can monitor his line in silhouette. He loosens the drag further so the reel sings, and when it does so a loon calls from the next lake over as if in response. The flute-like tone rises out of the darkness, falls like a wolf howl, and echoes among the trees. It is the classic wail that loons use to reach out to one another over long distances.

Something about loon calls can stir strong emotions in people as well as loons. Filmmakers often use recordings of them as spine chillers in their sound tracks to evoke a sense of foreboding, although they can also draw snickers from people who are familiar with them ("What is that poor bird doing in a desert?"). For many of us who love the North Country, however, the call of a loon is the haunting voice of wilderness itself.

It wasn't always that way. During the nineteenth century visitors and locals alike shot the big, black and white speckled birds on sight. Loons are hard to hit because they cruise low in the water and can dive to hide, a tempting challenge for target practice. Lacking the flavor of ducks and geese, fish-eating loons were not valued and therefore not "wasted" when killed on a whim. Jacklight hunters in particular hated them because their nocturnal calls sometimes startled deer who came down to the water's edge to drink at night, only to be dazzled by a torch and shot from a boat.

Writers who promoted Adirondack tourism at the time amplified anti-loon sentiments by describing their calls as "screams" or "weird laughter." The so-called screams and laughs of loons, however, are actually signals of distress, territorial aggression, or agitation, not of lunacy or humor.

How things have changed. Loons are now such desirable symbols of environmental quality that the presence of a nesting pair on a lake can boost the appraised value of waterfront property.

Little by little, Ted reels more line in than the fish drags out. I swing the canoe sideways to the Moon so I can more easily watch the exhausted animal break the surface. Ted reaches his free arm down into the chilly, black wetness and gently lifts. A huge trout rests quietly on his hand, half submerged, as if waiting to be unhooked.

In the old days, anglers often flaunted long stringers of trout or wall-mounted their largest, most evolutionarily fit catches as trophies, oblivious to the carnage they wrought on local fish communities. In years past both Ted and I have done much the same, but this trout will go free tonight. Ted doesn't even take a measurement before lowering his catch

Ted Mack with a brookie. *(photo by Curt Stager)*

back down and rocking the animal slowly back and forth to coax oxygen through the gills. I guess a length of roughly 2 feet (60 cm) before the fish slips out of sight with a liquid flick of the tail.

Empty-handed now but smiling, neither of us says a word. Tomorrow we'll brag to our colleagues without a body, a photo, or hard numbers to show for our evening on the lake. We know they will understand what happened here tonight, and approve.

Loon-shooting and trout-hoarding are unacceptable in the Adirondacks today. They reflected attitudes that were more common before ecology was a science and before our evolutionary connections to all life were as well documented as they are now. Rather than simply denouncing those outdated behaviors, however, we can also use them to measure how far we have come in a relatively short time.

Adirondack forests are generally in better shape than they were a century ago, having largely recovered from uncontrolled clear-cutting and accidental fires. The 6 million-acre (2.4 million ha) Adirondack State Park encloses a patchwork of private and public lands that supports a hundred small towns, some one hundred thirty thousand permanent residents, and nearly twice that number of seasonal residents. The core of the park is centered on forty-six peaks over 4,000 feet (1,220 m) in elevation, including mile-high Mount Marcy, the tallest in the state. Approximately three thousand lakes and ponds fleck the terrain, although nobody seems to agree on the actual number. Like the forests and mountains, many of them look much the same as they did centuries ago, and people still travel great distances to enjoy them, but conditions beneath their attractive surfaces are less reassuring.

When I first joined the faculty at Paul Smith's College in 1987, all I knew about the local lakes was that they were numerous and beautiful. It wasn't until I collected sediment cores from seemingly pristine lakes in the Saint Regis Canoe Area that I noticed something odd about them.

When an Adirondack lake turns over and mixes in spring and autumn, dissolved oxygen often rusts the iron-rich minerals in the bottom sediment. The rusting produces a pale oxidized layer atop darker sediments, a contrast that is strikingly obvious through the transparent walls of a core barrel. That is not what I found. In lake after lake, the youngest surface muds were much darker than the older, deeper ones, a possible sign of recent stagnation in the bottom waters.

That discovery led to others. I learned that many Adirondack lakes were being stocked with semidomesticated fish, most of whom were not expected to survive a full year. Others had been poisoned with rotenone or toxaphene to remove unwanted species. The darkening of the recent muds could reflect any or all of those legacies through the restructuring of food webs, decaying fish flesh, and other boosters of plankton productivity.

The deeper I looked into the stories of the lakes around me, the longer the list of problems grew. Road salt contaminated virtually every Adirondack water body within seepage distance of a town or highway. Some lakes had tons of lime dumped into them like force-fed antacids that only temporarily reversed the effects of acid rain. Others had been dammed in order to provide power or float logs to mills, which allowed their elevated surfaces to chew deeper into shorelines and release algae-stimulating nutrients.

Even much of the wildlife was not as it seemed. I learned from Ted that old-timers used to be able to tell what lake he had caught a brook trout in simply by looking at it. As he recalled it, each lake had its own distinctive variety, perhaps lighter or darker, slimmer or fatter, or more or less colorful than others. In contrast, most of the brook trout we encountered on our fishing trips were generic hatchery transplants. The rainbow trout and brown trout whom I also enjoyed catching were non-natives stocked for sport, and the splake were sterile hybrids who had been created by crossing lake trout with brookies. Even many of the beavers and bald eagles were immigrants, having descended from captives who were imported from western states after the native versions of their kind were decimated by trappers, random shootings, habitat loss, and pesticides.

One day while my paleoecology class was discussing that litany of woes, a student asked, "Aren't there any pristine lakes left at all?" Her question was rhetorical, but it sparked an idea. Rather than simply continue to document the depressing cases, why not also try to find something positive amid the gloom? Maybe there were still some true gems out there amid the costume jewelry of faux wilderness. An undisturbed lake would be a rare treasure, and from a scientific perspective it could be a valuable benchmark against which to measure environmental changes in other lakes as well as a point of reference for restoring damaged ecosystems.

The quest for an undisturbed Adirondack lake soon became an obsession for me, but it was not the first quest of its kind here. Thirteen years after Thoreau built his cabin at Walden Pond, Ralph Waldo Emerson spent several days camping beside Follensby Pond in the company of Swiss-born scientist Louis Agassiz, eight other prominent citizens of the greater Boston area, and a team of Adirondack guides. Their "Philosophers' Camp" sojourn became enshrined in history through Emerson's poem *The Adirondacs*, as well as essays and a painting, *The Philosophers' Camp in the Adirondacks,* by its organizer, William Stillman.

The Philosophers' Camp is often described as a transformational event in which elite luminaries of the arts and sciences came to the Adirondacks in order to ponder humanity's place in nature and blaze an intellectual trail that the modern environmental movement would later follow. In reality, it was more of a hunting and camping adventure, and its name was apparently coined by the guides, not their celebrity clients. Nonetheless, it was a high-water mark of sorts for Emerson and Agassiz. Follensby Pond was the closest Emerson would ever come to a real version of the idealized Eden of wild nature that he had made his reputation speaking and writing about. Agassiz was then at the peak of his fame for demonstrating that floods of glacial ice, not Noah's flood, had sculpted northern landscapes, but he would soon begin to fall from grace as a holdout against Darwinian evolution. He also believed in separate creations of the races, and at the time of the trip his writings on the subject were being used by slave owners to justify human bondage.

The Philosophers' Camp took place on the eve of seismic shifts in

Follensby Pond. *(photo by Curt Stager)*

American culture when wild frontiers were retreating and the Civil War was brewing. It offers a unique glimpse into past perceptions of wilderness, nature, and the human condition that still influence us today, and it also illustrates the kind of setting I would seek on my own quest.

It wasn't easy to visit the Adirondacks in 1858. That fact alone makes the Philosophers' Camp experience at Follensby Pond unattainable today. To awaken at dawn to the call of a loon on a lovely Adirondack lake can be a joy, but it is a more complex experience when that lake lies several days' hard travel from the nearest major city or medical facility. Panthers and wolves still worried visitors during the mid-nineteenth century, an infected cut or dysentery from untreated drinking water could be fatal, and there were few established trails to prevent inexperienced campers from becoming permanently lost if they wandered off alone. For most of us today, American wilderness is savored in relative comfort and safety. In 1858, it was more likely to kill you.

Stillman, who had already spent several summers in the Adirondacks, saw to it that the trip from Boston to Follensby would be as safe and enjoyable as possible. The difficulty in reaching the lake only added to its appeal. In early August the group disembarked from a steamer on the western shore of Lake Champlain, after which a stagecoach carried them the remaining 4 miles (6 km) to Keeseville where a welcoming party gathered. It was Agassiz who drew most of the crowd. He had recently become a darling of the American media, in part by rejecting an offer of a chairmanship at the Jardin des Plantes in Paris in order to accept a post at Harvard University. One member of the party held up a newspaper etching and cried, "Yes, it's him!" The admiring audience clustered around Agassiz and left Emerson and the others to their own devices.

Later, horse-drawn wagons hauled the men and their equipment over rough roads to Ausable Forks, then uphill through Black Brook. They probably stopped several hours later in Franklin Falls, where most visitors to that part of the uplands spent the night in a modest hotel. The next day the wagons followed the Saranac River to Martin's, a multistory luxury lodge just outside the sawmill village of Saranac Lake. On the third day a flotilla of small wooden boats, each rowed by a professional guide, shipped the travelers across Lower, Middle, and Upper Saranac Lakes to a more rustic rest house at Indian Carry. Day four saw the boats wagon-hauled to the Raquette River, where several more hours of rowing brought the group to Follensby. Nowadays, the Philosophers' four-day journey from Keeseville to the Raquette can be retraced on paved roads within less than two hours.

The Follensby excursion was later recorded in detail by Stillman's paints and prose, but Emerson's poetry was more evocative if less accurate. Emerson was not a scientist and did not seem to care much for technical details. "Empirical science is apt to cloud the sight," he once wrote, "and, by the very knowledge of functions and processes, to bereave the student of the manly contemplation of the whole." His poem had the group paddling north along the lake when the true route

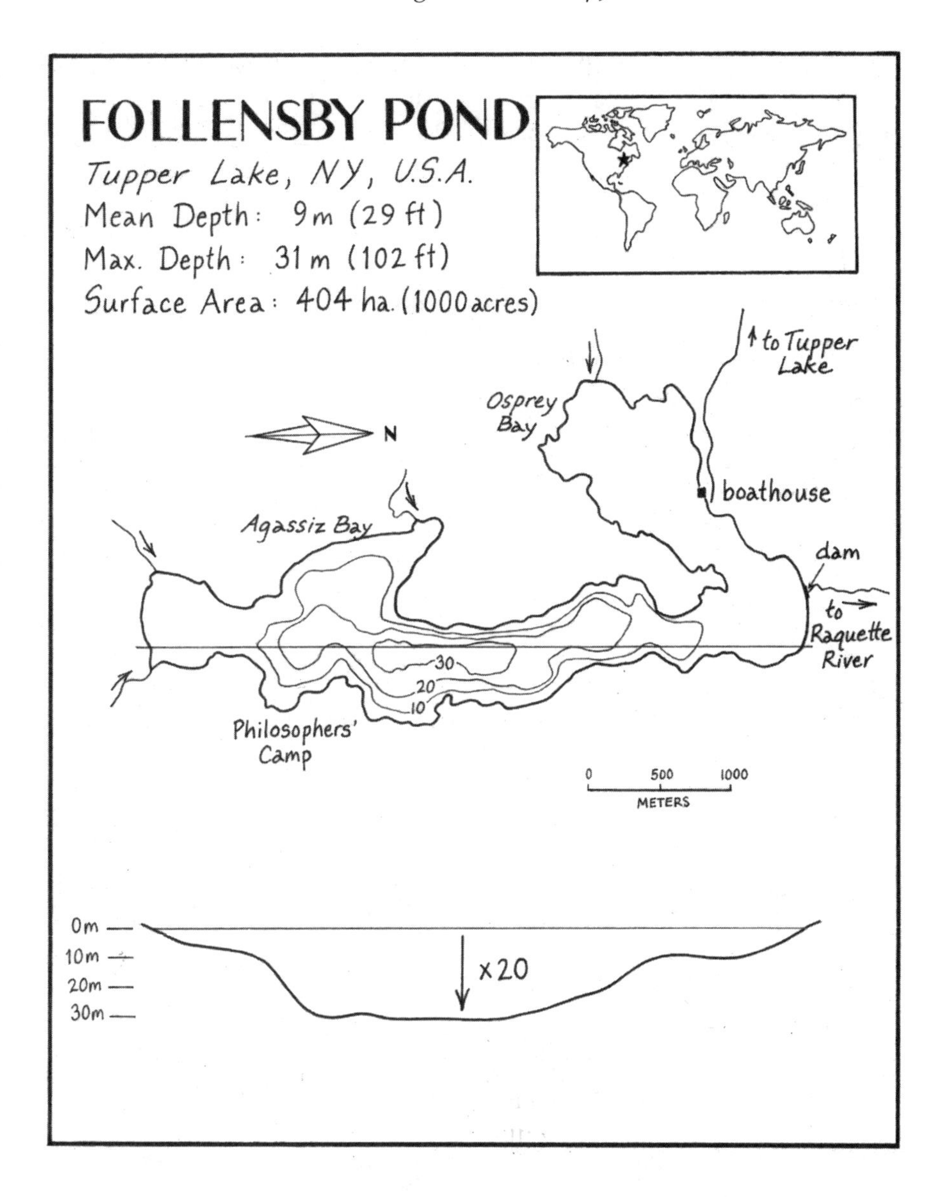

was southward, and it misidentified some of the animals and plants they encountered. Nonetheless, it vividly expressed his first and only direct, personal impressions of deep wilderness.

The entry to Follensby was an outlet stream that meandered through a lush wetland in squiggly loops, "two creeping miles of rushes, pads, and sponge, to Follansbee Water and the Lake of Loons." Stillman later wrote that Agassiz discovered a freshwater sponge there that was new to science, although it is more likely that it was simply new to Agassiz, whose expertise was in geology and fish rather than green crusts on submerged sticks. It probably belonged to the widely distributed species *Spongilla lacustris*.

Follensby Pond is shaped like a fishhook with its longer arm reaching nearly 3 miles (4.5 km) south from the outlet. The guides rowed their clients to a sheltered cove on the southeastern shore that Stillman had selected on a previous scouting trip. In Emerson's poem, they stepped ashore and "in the twilight of the forest noon" began to "wield the first axe these echoes ever heard." No matter that other Euro-Americans had previously hunted, fished, and camped beside the lake, as had indigenous peoples before them. No matter, either, that the town of Tupper Lake was close enough for group members to shop for supplies and receive news of the laying of the first undersea telegraph cable between Europe and North America. The thickly wooded shore was free of clearings and roads, nothing like Walden Pond back home in Concord. For sheltered city dwellers such as Emerson the feeling of isolation was delicious and flavored with just the right amount of danger.

Their sylvan "Camp Maple" lay beside a large boulder, a memento

William Stillman's painting of Camp Maple.
(courtesy of the Concord Museum, Concord, MA)

of the last ice age that Agassiz would surely have noticed and appreciated. Tall white pines and maples shaded bark-covered sleeping shelters that were furnished with pungent evergreen boughs for bedding. A shaded spring offered cool water to drink, and the guides set lines out in the lake for trout. At night "stars peeped through maple boughs which o'erhung, like a cloud, our camping fire," and luminescent fungi on fallen logs "like moonlight flecks, lit with phosphorescent crumbs the forest floor."

The beauty of the setting, however, did not prevent the men from enjoying it in ways that many of us would now object to. They drove deer with dogs and shot at them by jacklight at night. When a nesting osprey drew their attention, they urged a guide to climb its high pine roost and pilfer an egg. According to Emerson, poet James Russell Lowell also "watched for a chance to shoot the osprey, but he soared magnificently and would not alight."

"How went the hours?" Emerson's poem continued. "All day we swept the lake, searched every cove . . . whipping its rough surface for a trout; or bathers, diving from the rock at noon; challenging Echo by our guns and cries; or listening to the laughter of the loon; or, in the evening twilight's latest red, beholding the procession of the pines."

Agassiz kept busy by studying insects, dissecting deer, and weighing trout brains with Jeffries Wyman, an equally capable scientist. Although they would disagree on the relative merits of evolution and creationism, they worked well together and collected everything that moved, including dragonflies, salamanders, and an herbarium's worth of plants. "As water poured through the hollows of the hills to feed this wealth of lakes and rivulets," Emerson wrote, "so Nature shed all beauty lavishly from her redundant horn." A century and a half later, a BioBlitz survey sponsored by The Nature Conservancy also found rich biological diversity at Follensby despite the environmental carnage that followed in the wake of the Philosophers' Camp. The informal army of professional and lay naturalists identified more than five hundred species of animals, plants, fungi, and microbes.

Stillman later wrote that they had seemingly "got back into a not too greatly changed Eden." Near the end of the essay, however, he conceded that "our paradise was no Eden" after all. Tourists flooded the region in the decades following the Philosophers' Camp, drawn by glowing accounts such as Emerson's poem and images such as Stillman's painting. Unchecked timber harvesting, accidental wildfires, charcoal and potash production, iron mining, overhunting, and overfishing would soon turn much of the Adirondacks into a scarred wasteland. By the late 1800s, more than a million fresh-cut logs floated down the upper Hudson River over the course of a single year. Near the end of his life, Stillman returned to Follensby to write a retrospective article about it and was horrified to find the place desecrated. Campsites dotted the shore, "wretched dolts" had released trout-eating "pickerel" (more likely northern pike) into the Raquette River, and Camp Maple was so denuded by forest fires and axes that he could locate it only by its landmark boulder.

There is much to ponder in the story of the Philosophers' Camp, as the later writings of authors such as Christopher Shaw, James Schlett, and others attest. But what does it tell us about lakes and our relationships to the natural world?

The Philosophers' accounts illustrated the main criteria that most of us still require for a lake to be desirable. Above all, it must be pretty. For it to represent wilderness, as well, there must be no obvious signs of human presence or influence. As was the case at Camp Maple, however, the illusion of having it all to ourselves is often an acceptable substitute for reality. Follensby easily fit the bill in 1858, but it was the forest, mountains, and sky more than the lake that drew the most praise from Emerson and the others. The lake mainly reflected the setting, and most waters can do that just as easily.

Rarity also adds value, so the perfect wild lake must be difficult to reach and shown to be of exceptional quality and purity. Stillman boasted that the Follensby wilderness was larger and better than Thoreau's, and his criterion of exclusive personal use resembled a territorial defense of private hunting grounds. "This was not the solitude of Walden Pond," he wrote in an article for *The Century*, "where no bird, leaf or tree was ignorant of the daily footfall of idlers and curious, but a virgin forest, where the crack of our rifles reached no other human ear."

In reality, others could surely hear their guns because the lake and its surroundings were often used by local residents, but to Stillman and his colleagues the illusion of undefiled perfection was apparently sufficient.

As wonderful as Follensby was, however, it was not the epitome of all lakes for the Philosophers. Their ultimate Eden-like prize was, perhaps by definition, unreachable. Rumors of another, more remote lake just over a nearby hill made them crash about in the woods for a day in a vain effort to find it. Emerson called the elusive lake their "carbuncle" after a magical garnet in a story that Nathaniel Hawthorne had recently published. Seekers followed the gem's glow into the mountains only to be killed, blinded, or lost in fog. When a good-hearted young couple finally reached it, their willingness to leave it untouched saved them and, ironically, caused the carbuncle to fade.

To FIND A TRUE wilderness lake, one must first define what one is looking for. By the simple metric of being unaltered by humans, pristine wilderness is far more abundant than most of us realize. Species diversity on the microscopic scale is still thriving worldwide, and most tiny organisms go about their business all around and even within us, both undetected by us and with no clue that we exist, either.

Wilderness also dominates the universe on the cosmic scale as it always has. NASA scientists have recently estimated that there are as many as 10 billion Earth-like planets in our Milky Way galaxy alone, and those planets surely support many billions of lakes that humans will never affect in any way. If you crave untouched nature for its own sake, and lots of it, then your wish is granted. To gaze upon unfathomable depths of truly wild wilderness, simply travel far enough from bright lights to view the Milky Way on a clear night.

To be desirable in practical terms, however, a pristine lake must also be at least somewhat accessible. Earth does support some untouched lakes, but most of them are entombed in deep caverns or trapped under thick polar ice sheets. Every other lake that is exposed to the modern atmosphere, from Alaska's frigid North Slope to the rain-soaked tip of Tierra del Fuego, contains radioactive fallout from nuclear bomb tests, carbon from fossil fuel combustion, or other human-generated substances.

My students and I decided that the most useful approach to the quest was to step back from impossible perfection to the "good enough" position of minimal impact. We also set a specific time frame within which to judge the status of our candidates. They were to be essentially the same as they were in AD 1800 when air pollution was not yet globally significant and Euro-American settlers had yet to arrive in force.

With thousands of Adirondack lakes to choose from, we rejected any that bore obvious signs of human influence. That eliminated the largest ones because all of them had been stocked with hatchery fish, opened to motorcraft, dammed, or decorated with homes, campgrounds, or towns. We also consulted a database compiled by the Adirondack Lakes Survey Corporation (ALSC), who had sampled more than a thousand lakes during the 1980s to take stock of pollution, invasive species, and other environmental threats. Any lake with low pH was rejected as a likely victim of acid rain along with all waters containing nonnative species. A DEC fisheries officer provided a list of roughly one hundred lakes that the state had poisoned with rotenone or toxaphene, and other sources showed which lakes were limed or contaminated by road salt. The winnowing process left us with very few candidates to investigate, and I began to wonder if we would find any suitable lakes at all.

Meanwhile, ALSC research manager Karen Roy and I focused the concept by giving it a name. Borrowing a term that was then being used by the DEC to distinguish native heritage brook trout from run-of-the-mill hatchery fish, we called our quarry "heritage lakes." Word of the project spread through the local community, and people began to suggest possible heritage lakes that were not in the ALSC database.

Our first break came during the summer of 1996 when ecologist Dick Sage invited me to visit the Huntington Wildlife Forest in the central Adirondacks, a few minutes' drive from the town of Newcomb, population 436. The 15,000-acre (6,070 ha) tract had been donated to the State University of New York (SUNY) by the philanthropists Archer and Anna Huntington during the 1930s, and SUNY faculty and students had conducted a great deal of research on its woods, waters, and wildlife. When Dick learned of my quest he suggested that one of the lakes on the property might be worth investigating, as well.

Wolf Lake is a source of the Hudson River, nearly 1 mile (1.5 km) long, about 35 feet (11 m) deep, and accessible only by a gated dirt road. Written records showed that it did not acidify as many other Adirondack lakes had during the mid-twentieth century, nor had it ever been dammed, stocked, limed, rotenoned, or otherwise "improved." So far, so good.

On the twenty-minute drive along the narrow, winding forest road to Wolf Lake, Dick pointed out signs of other research projects that were currently under way. A car-sized wooden crate beside the route was a live-trap for black bears. A deer who sauntered across the road in front of us wore a yellow identification tag on one ear. Dick pulled over and led me to a large square of chain-link fence that seemed to be stuffed with salad. It was a deer exclusion cage. Adirondack forests are

The author and students at Wolf Lake. *(photo by Brendan Wiltse)*

tended by four-legged gardeners, and without their constant nibbling the open stands of maple and beech could become jungles.

We turned off the main track and parked in a grassy clearing where shimmering flecks of blue sifted through a lakeside grove of fir and spruce. A footpath led down a short slope to a rickety log cabin, former shelter for researchers that was now abandoned to the porcupines. A few feet past the cabin the path ended on a narrow beach of creamy white sand. Green forest-cloaked hills to the left and right cradled a gorgeous lake upon which a pair of loons cruised slowly, then vanished in unison as they dove for dinner.

"Welcome to Wolf Lake," Dick said with a smile.

The following year, I organized a team of students to check for invasive species. It was a chilly, wet afternoon in October when I introduced Thom, Corey, and John to Wolf Lake. The water was as gray as the sky, the hills surrounding us were painted with maple-red and birch-yellow, and the damp air carried the sweet scents of fallen leaves and the moist duff of the forest floor. We stretched a trap-net several dozen feet out from shore, leaving a pocket at the far end to corral shore-cruising fish. The next morning the pocket was full.

Most of the fish squirming in our net were *Catostomus* suckers up to 1 foot (30 cm) long with silvery flanks and downturned lips. Scooping the fish back into the lake one by one, we also uncovered spiny sunfish, fat little chubs, bewhiskered bullheads, and a few small brook trout. To the average person those fish might seem unremarkable, but to us they were a thrilling discovery. Every species in the net was a native, representing the first intact fish community I had ever seen in an Adirondack lake. Our catch also challenged some widely held beliefs about the management of trout waters in these mountains.

Travel writers of the nineteenth century depicted Adirondack lakes as teeming with brook trout and promoted the region as a sportsman's paradise. The selective reporting created an enduring image of wild lakes as highly productive trout monocultures. Later, when trout became scarcer under heavy fishing pressure, other kinds of fish were often blamed even where selective angling had simply given overlooked species a competitive edge. Suckers and bullheads were, and often still are, seen as pesky

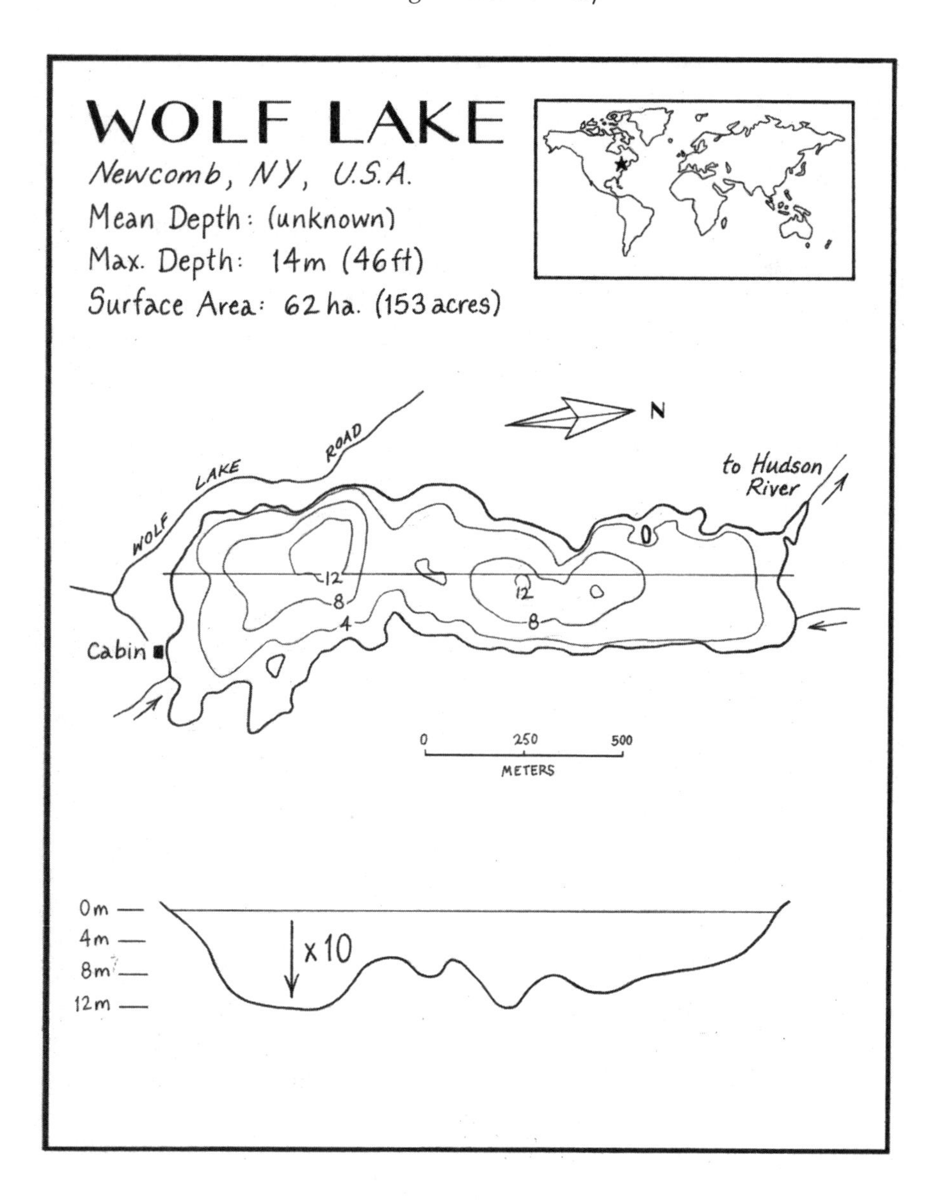

eaters of trout eggs or competitors for trout food, vermin to be eliminated in order to provide more desirable angling. Wolf Lake, however, had shown us what a real nineteenth-century Adirondack fish community looked like, a diverse mix of species as would be expected when a lake is a wild ecosystem rather than a fish farm.

The brook trout we found in the net were not large enough to decorate a sportsman's wall, but they likely represented an as-yet-undocumented heritage strain, one that was specifically adapted to life in Wolf Lake. Like the other heritage trout of the Adirondacks, those fish were reservoirs of genetic diversity who might later be able to help their species survive in an unstable future. Biologists from Cornell University have recently demonstrated that high water temperatures make it difficult for Adirondack brook trout to reproduce, and possible heat-tolerant local strains may become more important here as the world warms.

We would later learn that there was also more to the suckers than we realized at first. According to DEC biologist Doug Carlson, some of them might have belonged to a unique species found only in the Adirondacks. Previous netting studies on the lake had recorded "dwarf white suckers" who spawned shortly after regular suckers (*Catostomus commersonii*) in early summer. Careful examinations of similar fish from other Adirondack lakes have since documented an endemic species, *C. utawana*.

All of this precious biodiversity—the intact fish community, the heritage trout, the suckers—would have been lost if Wolf Lake had been treated in the manner that most sizeable Adirondack lakes are. From the perspective of many fisheries managers the lake is infested with trash fish and therefore ripe for reclamation and stocking. Even if the DEC chose to protect it instead, Wolf Lake might still be vulnerable to the sad fate of Little Tupper Lake.

In 1998 a private Adirondack estate that had once been the size of Manhattan was sold to the state of New York, and Little Tupper Lake was thereby opened to public access. When the DEC prohibited the use of live bait in the lake in order to protect the lake's heritage trout from invasive competitors and predators, it aroused the ire of a local fisherman. A mutual friend told me that this fellow, a resident of nearby Tupper Lake village, secretly released trout-eating bass into the lake because he resented being told by the government how he was allowed to fish.

With no recognition of heritage lakes in the management policies of the DEC, and with the selfish behavior of a single individual being capable of undoing the work of thousands of years of ecology and evolution, it seemed miraculous to find Wolf Lake's fish population in such an undisturbed state. The miracle that saved it, however, was not an act of God but a gate.

We had confirmed the fish community's intact status, but we had yet to do the same for the lake itself. Written records covered only the previous seven decades, not enough to tell us what Wolf Lake had been like during the early nineteenth century. My student Thom Sanger accepted the challenge by analyzing a sediment core from the deepest part of the lake for his senior thesis.

Thom worked his way down through the core centimeter by centimeter, identifying the diatoms in the sediments. Every sample contained the same list of taxa, mostly the starbursts of planktonic *Cyclotella* and *Discostella* with buckshot-blasted cylinders of *Aulacoseira*, as well as lemon-shaped *Achnanthes* and other diatoms who thrived on the bottom beneath clear waters. Lead-210 dating showed that the oldest layers at the base of the core were deposited during the late 1700s to early 1800s. We had found our first heritage lake at last.

What does one do with a heritage lake once it is discovered? Apart from protecting it, one does nothing but appreciate it and listen to what it tells us on its own terms. The staff of the Huntington Wildlife Forest are very careful to prevent avoidable human influences on Wolf Lake. Access is strictly regulated, no angling is allowed, and the only boats permitted to touch the water are canoes and rowboats that are left on shore for researchers to use, all important rules that help to keep hitchhiker species from entering the lake. The point is not to manipulate it but to learn from it in ways that help us to better understand other lakes, as well.

With support from the National Science Foundation, I later collected more cores with my students and several colleagues from Queen's

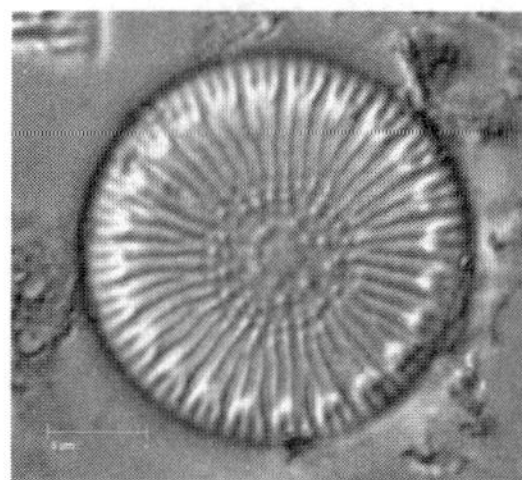

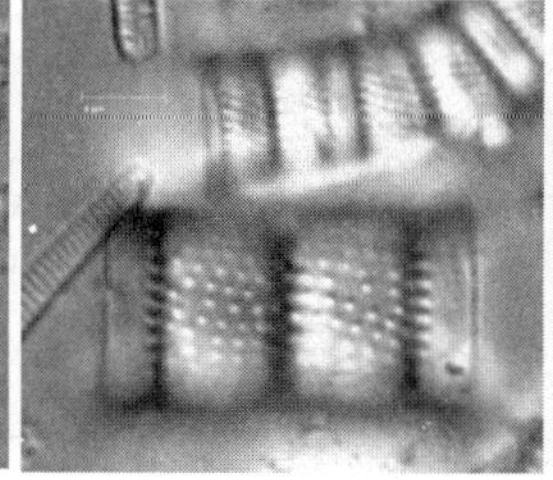

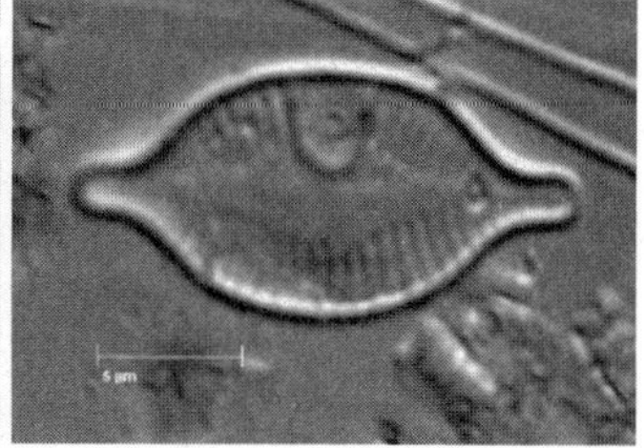

Diatoms from Wolf Lake sediment cores. Left: Planktonic *Cyclotella bodanica*. Center: Semi-planktonic *Aulacoseira distans* and *A. lirata*. Right: Bottom-dwelling *Achnanthes*. *(photos by Curt Stager)*

University, Ontario, to probe deeper into the sediment record. Minor fluctuations in the relative abundances of certain species told of prolonged droughts roughly 900 to 1,200 years ago, and a slight increase in the planktonic forms since the mid-nineteenth century seemed to reflect small-scale logging operations in the watershed. Despite those modest changes, our longest cores demonstrated that the overall tally of diatom species in Wolf Lake had remained unchanged for thousands of years.

Our quest for a heritage lake was a success, but it soon proved poignant. We didn't realize when we found our long-sought lake that it was already beginning to slip away from us. The recent increase of planktonic diatoms meant that nutrient inputs from previous logging activity might have helped to set the stage for water quality problems that are only now becoming evident.

In May 2015, Wolf Lake turned sickly green. Water that had once been clear enough to reveal a white Secchi disc when it was lowered 27 feet (8 m) below the surface now obscured it within the first 6 feet or so (2 m). Back in my lab, I slipped a sample of lake water under the microscope and saw tiny green specks of *Chroococcus*, a kind of cyanobacterium that had not been reported there in such abundance before. At first I thought it might simply reflect an unusually vigorous response to the spring overturn, but the lake remained murky through the summer and into the fall. Had some invasive fish colonized Wolf Lake and disrupted the food web by grazing down zooplankton that might otherwise keep the phytoplankton in check?

I emailed SUNY professor Kim Schulz, who had been monitoring the fish community of Wolf Lake for several years. She assured me that there were no new fish species in her nets. However, a recent study led by SUNY ecologist Colin Beier also demonstrated that Wolf Lake's ice cover season is now several weeks shorter than it was during the 1970s. The unexpected bloom could therefore reflect something more widespread and troubling than the local arrival of nonnative fish.

Lakes all over the world are turning color because we are changing the atmosphere that sustains them. In the Canadian Arctic, formerly frozen lakes are now thawing in summer and allowing sun-loving plankton to thrive in them for the first time. In tropical Africa, lakes are becoming warmer and more strongly stratified, which often favors cyanobacteria. In the American Rockies formerly clear, dilute alpine lakes are being overfertilized by airborne nutrients from cities and farms, and increasingly stratified, stagnant waters are releasing phosphorus from bottom muds.

The lake was still green in September 2016, when a routine plankton haul opened yet another chapter in the story of Wolf Lake. SUNY scientists were shocked to discover something strange floating in their samples—clear, colorless discs the size of a thumbnail that pulsated in the murky broth. They were freshwater jellyfish.

The newcomers belonged to a species (*Craspedacusta sowerbii*) that is native to Asia but has quietly spread throughout North America during the last century. The jellies are most often dispersed by contaminated boats or the emptying of aquaria and garden pools into lakes and streams. At Wolf Lake, they more likely arrived as larval stowaways on the legs or feathers of migratory waterfowl. Some of our Adirondack loons might now be unwitting agents in the dispersal of invasive organisms such as these, thereby contributing to the disappearance of the kind of wilderness they have come to represent. Although *C. sowerbii* jellyfish are harmless to humans, they feed on zooplankton, and their recent arrival might therefore have amplified the greening of Wolf Lake through top-down trophic effects.

At the time of this writing, we still don't know exactly what changed Wolf Lake's plankton community or if the clouding of its waters is permanent. Regardless, it is clear that even remote and heavily protected waters are no longer immune to human influence and that Wolf Lake's heritage status is now history.

I deeply regret losing our first heritage lake, but as a scientist I also know that there is little hope of preserving most wild waters in their preindustrial states forever, and that instability itself is not necessarily unnatural. Lakes have always changed with climates and the migration and evolution of species, and the question of how best to respond to environmental change in protected places is an area of active debate.

Restoration of altered ecosystems was one of Aldo Leopold's key concepts in *A Sand County Almanac*, and many people see it as a way to make amends for past transgressions against the natural world. Critics say that it may be futile now that it has become a running battle with climate change, and that such efforts rarely restore the entire living community anyway because we rarely have the resources to do so or full historical accounts to refer to. For them, restoration is little more than gardening, a reflection of antiquated myths of Eden and a stable balance of nature that never actually existed.

In the case of Wolf Lake, forcing it to remain unchanged forever would have run counter to the concept of heritage lakes, which are not to be manipulated but monitored. Instead, the scientific value of Wolf Lake has evolved along with the ecosystem itself. It is now not so much a heritage lake as a sentinel that warns of impending changes in other Adirondack waters, like a canary in a coal mine.

Fortunately, many farsighted and well-informed people recognize that human-driven environmental change is already under way and are designing management plans for wild places in an unstable world. The best of those plans anticipate and prepare for an evolving future with the aid of first-rate science and an understanding that we play important roles in most ecosystems. In some cases, as with acid rain and climate legislation, federal oversight may be required when the priorities of states or industries work against the public interest. In other cases, state or private organizations can better address environmental problems in creative ways when federal involvement is inappropriate or absent, which brings us back to Follensby Pond.

Follensby may be a greatly changed Eden but it has regained much of its wild appeal since Stillman's sad return during the late 1800s. The forests around it have grown back, and although the lake now stands slightly higher because a dam was built on the outlet, it looks much like it did to the residents of Camp Maple.

After becoming a private estate for several decades, the lake and the land around it went up for sale in 2008. Recognizing its historic significance and exceptional beauty, the Adirondack chapter of The Nature Conservancy purchased the 14,600-acre (5,910 ha) parcel for $16 million, hoping thereby to give the state of New York time to rally funds to buy it from them. Shrinking budgets stalled the deal with the state, however, leaving the Conservancy with a massive expense on its books and a unique lake on its hands, the largest privately owned water body in the northeastern United States.

For now, access to Follensby is limited to researchers and Nature Conservancy staff, and the entrance road from Tupper Lake village is gated. However, conservation program director Dirk Bryant granted me permission to explore the famed lake and collect a sediment core from it. Concerned by the recent changes at Wolf Lake, I wanted to see if a similar transformation was also under way at Follensby.

ON A WARM, sunny day in June 2016, Nature Conservancy researcher and writer Mary Thill leads me to a wooden boathouse on the northern shore of Follensby Pond. Caretaker Tom Lake is waiting for us there. He is a longtime resident of the nearby town and a dedicated guardian who knows Follensby, the surrounding woods, and his neighbors well. He makes sure that the gate remains locked and that the few boaters who try to navigate the marshy route from the Raquette River depart soon and leave no trace. If Follensby is once again a wilderness Eden, then Tom is the fierce angel of Genesis who protects it.

Mary opens a contour map of the lake that was made by my former student Brendan Wiltse, who recently completed a detailed echo-sounding survey for the Conservancy. Follensby is to be more than just another pretty wilderness park. It will also be an ark, a refuge for cold-water fish in a warming world. The centerpieces of the refuge will be the lake trout, still as abundant here as they were when Emerson and Agassiz "swept" these waters.

Mary summarizes for me the plight of North American lake trout while Tom helps us to prepare a canoe for travel. They require well-oxygenated water in the 50–54°F (10–12°C) range, which restricts them to northern lakes and keeps them deep in summer. In the Adirondacks, the combination of overfishing, invasive species, rotenone poisonings, water quality problems, and climate change could become an existential threat to lake trout unless appropriate measures are taken to protect them for the long term. That's where Follensby comes in.

Fewer than 1 percent of Adirondack lakes still contain lake trout, and only half of those support self-sustaining populations without help from stocking programs. Follensby's population is therefore unusual in this region, a largely unexploited community akin to an old-growth stand of ancient trees. The Conservancy has also documented a healthy population of cisco or "lake herring" that lake trout love to eat. Perhaps most important to the future of the trout, however, is the physical form of the lake, as illustrated by Brendan's map. Its large, deep, bathtub-shaped main basin offers plenty of room for trout to escape the summer heat until the fall overturn recharges its oxygen supply.

Mary and I take our seats in the canoe and push off from the dock. Our plan is to see what remains of Camp Maple, but first I want to visit the stream that the Philosophers followed to reach the lake. A ten-minute paddle brings us to the splash-dam that was built by a timber company during the late nineteenth century. It raised the level of the lake by several feet so logs from the surrounding forest could be floated to Tupper Lake for processing. The removable boards that controlled the flushing are now gone, but the concrete base still keeps the level slightly higher than before.

Beyond the dam, a marsh brimming with sedges, shrubs, and wildflowers stretches northward to the Raquette. We set off on the narrow, winding stream, half-paddling and half-pushing against the sandy bottom. I imagine Agassiz pulling scraggly green sponges aboard, but I find none myself. June is too early for sponges to have grown back after their winter demise. No matter. I am easily distracted from thoughts of 1858 by the warmth of the Sun on my face, the faint clacking of dragonfly wings, and the scents of marsh blossoms.

Emerson would have approved, I'm sure. In his essay *Nature*, he railed against tendencies to focus too much on the past, meaning the old world of Europe, when judging the quality of life in what was to him a new America. "The sun shines today also," he wrote, and I concur. The world today is quite different from that which Emerson and Agassiz

Paddling upstream to Follensby Pond. *(photo by Curt Stager)*

knew, in part because the intervening century and a half of science has revealed so much more about it.

We return to the beach and find it streaked with black particles that sparkle in the bright sunlight. I kneel for a closer look. Waves have sifted heavy flecks of iron ore from the quartz and feldspars and smeared them across the sand in dark, parallel stripes. One deposit is so rich that I can scoop both hands full and shape the damp mass into a soft, weighty cannonball. Enterprising Adirondackers of the nineteenth century sometimes mined the beaches of local lakes for such ore, but the Philosophers apparently made no note of it. Even if they had, they would not have recognized what modern science reveals here, far down at the small end of the size scale.

Rachel Carson once wrote that "in every grain of sand there is the story of the earth," and it is certainly true on this beach. The warm grains between my fingers are the powdered bones of long-vanished mountains that decomposed, fleck by fleck, under the slow erosive forces of rain and frost. They are, quite literally, the sands of time embodied in a beach. Feeling the heat of the Sun in them, however, also helps me to look even more deeply into their stories.

Iron atoms form in the hearts of massive stars, then cause them to explode and scatter the elements of potential life into the cosmos. The tiny star-killers on this beach helped to produce the primordial matter of the lake, air, and life forms that surround me as well as my own body. On the atomic level, the Earth and Sun are not our mother and father as some myths suggest but our siblings, offspring of the same parent star that birthed the solar system billions of years ago.

Mary and I walk to the water's edge to rinse the grit from our hands. I wade shin-deep and bend forward to sweep my fingers through the reflections. When the ripples settle I notice something lying on the bottom and pick it up. It is a flake of chipped flint.

My sense of the human history of this lake now deepens, as well. That little brown flake reminds me that people lived here long before the Philosophers' Camp. An ancient projectile point that was recently found on the shore of Tupper Lake tells of caribou hunters who lived nearby when the newly deglaciated Adirondacks were still covered with tundra more than ten thousand years ago. Human beings are not intruders on this forested landscape, and the unpeopled American wilderness that Emerson and his companions romanticized was a mere fantasy. The ancestral roots of humanity run deeper here than those of the trees.

We push off again and paddle south toward Camp Maple. The surface of the lake is so flawless that the water in front of us seems to vanish, and I feel as though I am suspended between mirrored wells of time. Agassiz saw the world as a young creation, a belief that confined him to a pool of history so shallow that it left no room for evolutionary connections in the global kinship of life. I imagine that he would have reveled in those connections if his worldview had allowed him to recognize them.

We enter a broad bay where I lower my core sampler overboard. I will later find little evidence of change in the sediments under my microscope apart from a very recent increase of chrysophyte algae much like the one that has occurred at Walden Pond, a hint of possible changes to come but nothing too serious yet. For now, though, I am thinking instead of the trout who may be watching from below. Melissa Lenker, a biologist from McGill University who has been studying them, recently provided the Conservancy with a figure that I have never encountered before: the total number of fish in a lake. She estimates that at least two thousand adult lake trout live in Follensby Pond, and that roughly 100 of them are more than 30 inches (76 cm) long.

Follensby is an angler's dream, but we will not bother any fish today. Slow-growing lake trout are extremely sensitive to human predation, and the Conservancy's goal is for them to be self-sustaining far into the future. The staff have therefore decided that it is best to await a well-researched plan for low-impact use before opening the lake to angling, and they refer to the trout population as a "community" rather than a "fishery." The care being taken on behalf of this lake is exemplary, and it makes me wonder if the DEC could bring similar resources and enlightened perspectives to the management of Follensby Pond were it to take ownership of it.

Lenker's research suggests that even catch-and-release fishing could decimate Follensby's trout if not carefully monitored and regulated. The clear water contains relatively little plankton to support the food web, and the trout take more than a decade to reach breeding age on the limited fare available here. Furthermore, as many as 15 percent of hooked fish die of stress and trauma even if they are released, particularly when they are caught at great depths. Hauling a laker up from deep water terrorizes and tires the animal, risks damaging the mouth and gills, and causes the gas-filled swim bladder to distend or even burst. It can also overheat a cold-loving fish in warm surface waters, the very problem that the Conservancy hopes to avoid by preserving Follensby. Protecting the lake and its trout community is easier than trying to restore it once it is disrupted, and Lenker's analyses show that it could take many decades for the population to recover if it collapses.

My own choice to refrain, however, is not made only because of the angling ban. It is a personal position that has developed gradually since my moonlit encounter with Ted Mack's lake trout on Little Long Pond. The more I realize that fish are unique individuals who suffer when hooked and dragged to a boat, the less I enjoy angling. Rather than suppress my empathy in an activity that is no longer necessary for my survival, I prefer to develop it as a part of my evolutionary inheritance as a human being that is of value in the modern world. My decision has therefore been made not only because of what angling does to a fish but also because of what it does to me.

We now head for the cove on the southeastern shore where the Philosophers made their camp. Tom told us how to find Camp Maple, but it won't be easy. We are to watch for a small spring and a big boulder, but the original landmark trees are long gone and the former landing beach has been submerged by the dam.

We tie up to a cedar root where a line of rocks leads into deep water, a potentially inviting series of resting and diving platforms for bathers. The rough ground rises to a flat spot where tall maples, beeches, and birches shade the forest floor. A brooklet gurgles nearby beneath a canopy of ferns, perhaps the one that Emerson and Agassiz drank from. We also find several large boulders, any of which could pass for the rock

Glacial boulder at the former site of Camp Maple. *(photo by Curt Stager)*

in Stillman's painting. We may never know which one was the original. I have watched master painters at work, and I know that details are sometimes sacrificed in an artist's rendition of reality. I convince myself that I have touched the real Camp Maple boulder by touching all of them.

Our goal is not to relive the Philosophers' Camp but simply to savor and learn from it. The regenerated forest is too young to have shaded Camp Maple, and today's lake is now filled with more recent raindrops. We also know more about this place than the Philosophers did, and we can see farther into its future than they could, as well.

Agassiz was rightly famous for documenting the glacial story behind Follensby's boulders, but many scientists now envision a shocking sequel to it. Heat-trapping carbon dioxide from our fossil fuel emissions will likely contaminate the atmosphere for tens of thousands of years to come. Natural cycles in the tilt, wobble, and orbit of the Earth would normally trigger another glaciation fifty thousand years from now, but the lingering greenhouse gases that we have already released could warm the future atmosphere just enough to counteract the cooling. Since Emerson penned ideas that seemed to set humanity apart from the natural world, we have become a force of nature powerful enough to prevent the next ice age. If we burn all remaining fossil fuel reserves during the next century or so, the glacial boulders of Follensby might not see another ice sheet for half a million years.

We return to the canoe and open our lunch bags. Mary checks her phone and barely gets a signal. Ah, wilderness! Nowadays any place that lacks cell coverage feels like the far end of the Earth. Like the more or less invisible connections that evolution, ecology, and atomic elements bring to our lives, electronic connections now potentially link us to every other person on the planet. Even here where news of the trans-Atlantic cable sparked such enthusiasm in 1858, we easily take for granted the global web of communication and the ever-present eyes of satellites beyond the clouds. Much more of the world is revealed to us as a result of that technology, but it also poses new risks to which we have not yet fully adapted, including that of mistaking virtual reality for actual reality.

I recently enlisted some of those orbiting eyes to help me do something that the Philosophers tried to do and failed. I opened Google Earth on my laptop and zoomed in on a wooded hill near this spot where the men of Camp Maple went in search of their carbuncle lake. I was tempted to peek over that same hill through my computer screen and hunt for the hidden pool in my own way. Would it be cheating, I wondered? Certainly. Pursuing their dream lake from my living room would be like fishing for trophy trout in a bucket. Its greatest value, like that of the Edens in our imaginations, may be as an ideal that we admire but never fully attain. That is why I will not tell you what I found on that screen and will leave it up to you to decide whether or not to seek the Philosophers' lake yourself *in silico* if not on the ground.

The Sun hangs low over the trees, and it is time for us to leave. The weight of still water pulls Follensby's surface flat and glassy, and everything above it is soaked with soft, golden light. Backlit midges glow against dark reflections of the forest while dragonflies dart among them like fiery sparks. Where the gently wavering images of woods and sky meet, the mosaic of shadow and light resembles the checkered pattern on a loon's back. As we glide forward to the rhythmic murmur and drip of paddles, the warm air carries a pleasant but unusual cocktail of scents. I notice the fresh smell of balsam fir amid earthier tones of crumbling leaf litter, but also an additional sweetness that the lake releases, perhaps from the plankton that floats beneath us.

We pause beside a large red ball that hangs like a planet in the mirrored sky. The buoy supports a monitoring station placed here on the deepest sector of the lake by a researcher for The Nature Conservancy who is studying the temperature and oxygen content of the water column. Sensors dangle from the rope below it, and the combination of clear water and reflections fools my eyes so the whole contraption seems to float in thin air. It reveals real depths beneath that attractive but deceptive sky, and it reminds me of how science exposes illusions of our own making.

If romantic visions of wilderness that seemingly separated us from nature fade in the presence of science, then I say good riddance. Rather than mourn the supposed loss of a mythical paradise, I favor Emerson's observation that "the sun shines today also." Like his time, ours is one of great uncertainty but it is also an age of brilliant, transformative discoveries and reasons for both caution and hope.

We are beginning to recognize how interconnected we are with the natural world, not only in some vague figurative sense but in empirically demonstrable ways, both through our common ancestry and through the shared pool of elements from which we are made. And we are learning that, although life itself will continue with or without us, it may not be in forms that we prefer unless we become more responsible and well-informed stewards of our home planet.

We protect and study lakes for practical reasons but also because we love and learn from them as our forebears did for reasons of their own. With honorable intentions guided by good science and good taste, perhaps we can leave them in good enough shape for future generations to love and learn from under their own shining suns, too.

(photo by Curt Stager)

GLOSSARY

Acid rain: Rain that is contaminated by acidic pollutants such as the oxides of nitrogen (largely from motor vehicle exhaust) or sulfur (largely from coal-fired power plants).

Algae (*AL-gee*): Plant-like organisms belonging to the kingdom Protista, most of whom are green or brown, aquatic, and microscopic.

Anthropocene (*AN-thro-po-seen or an-THROP-o-seen*): An as yet unofficial but widely used name for the present epoch of geologic time in which humans have become a prominent force of nature; often translated as "The Age of Humans."

Autotroph (*AW-toe-troaf*): An organism who obtains energy from sunlight or inorganic chemicals rather than by consuming other living organisms. Examples include plants and algae.

Bloom: Prolific growth of planktonic algae or cyanobacteria in a water body who become so numerous that they may change the color of the water they live in.

Brownian motion: Random, heat-driven jiggling of microscopic particles caused by collisions with the smaller water molecules that surround them.

Chrysophyte (*CRISE-o-fite*): Golden-brown, single-celled algae who are often covered with glassy scales and able to swim with whiplike flagellae. Includes members of the genera *Synura* and *Mallomonas.*

Cichlid (*SICK-lid*): A member of the freshwater fish family Cichlidae.

Cladoceran (***kla-DOSS-er-an***)**:** A microscopic, free-swimming crustacean resembling a shrimp.

Copepod (***KOE-pe-pod***)**:** A microscopic, free-swimming crustacean resembling a shrimp. Usually more elongated than cladocerans.

Cyanobacteria (***sy-ANN-oe bak-TEER-i-ah***)**:** Greenish bacteria who resemble plants or algae. Formerly, and incorrectly, known as "blue-green algae."

Diatom (***DAI-a-tom***)**:** Golden-brown, single-celled algae with glassy shells.

El Niño (***el NEEN-yo***)**:** A semicyclic climate disturbance related to changing winds and ocean currents in the tropical Pacific Ocean, which is often accompanied by weather disruptions in diverse locations worldwide.

Endemic (***en-DEM-ik***)**:** Unique to a particular location. When applied to a particular species, the term typically implies that the species originated in that location.

Epilimnion (***ep-ee-LIM-nee-on***)**:** The upper layer of a thermally stratified lake, typically warmer and less dense than the layers below it.

Eutrophic (***yoo-TROE-fik***)**:** A habitat in which living organisms are extremely abundant and prolific. In reference to lakes, it usually means that the water is unusually rich in plankton and/or plant growth.

Halobacteria (***HAY-loe bak-TEER-i-ah***)**:** Bacteria who thrive in extremely salty environments.

Heterotroph (***HET-er-o-troef***)**:** An organism who consumes other organisms.

Hydrogen bond: A weak bond between molecules that involves the attraction between a positively charged hydrogen atom on one molecule and a negatively charged atom on another molecule.

Hypolimnion (***high-poe-LIM-nee-on***)**:** The lowermost layer of a thermally stratified lake, typically cooler and denser than the layers above it.

Internal loading: The release of nutritional elements such as nitrogen or phosphorus from the bottom sediments of a lake, typically when dissolved oxygen is scarce due to extreme depth or eutrophication.

Kin selection: An evolutionary process by which organisms who sacrifice resources or risk danger on behalf of relatives still tend to pass their genetic traits on to future generations by way of those relatives. Examples include warning calls and parental care of young.

Lake: A landlocked body of water with no direct connection to an ocean. Can be fresh or salty. Among limnologists there is no formal distinction between a "lake" and a "pond," nor is common public usage of the two terms consistent. However, a water body is more likely to be referred to as a pond rather than a lake if it is relatively small and lacks a surface outlet.

Limnology (*lim-NOLL-o-gee*): The study of lakes.

Loch (*LOK*): A traditional Scottish term for "lake."

Macroevolution: The origin of new species through evolution.

Mbuna (*mm-BOO-nah*): Local traditional name for cichlid fish who live in rocky habitats in Lake Malawi, East Africa.

Mesotrophic (*MEEZ-o-TROE-fik*): A habitat in which plankton and/or plants are moderately abundant and prolific, but less so than in a eutrophic habitat.

Microevolution: The origin of new genetic varieties through evolution without producing new species.

Natural selection: An evolutionary process by which certain varieties of organisms become more likely than others to survive and pass their genetic traits on to future generations.

Nutrient: A dietary substance that is necessary to build and/or maintain a living body. In lakes, the two nutrients that are most often the focus of interest are the atomic elements phosphorus and nitrogen, which can occur in various molecular forms, including phosphate, nitrate, and ammonium.

Odonate (*OH-do-nayt*): Member of the insect order Odonata, which includes dragonflies and damselflies.

Oligotrophic (*OLL-ig-o-TROE-fik*): A habitat in which plankton and/or plants are not very numerous or prolific, much less biologically productive than a eutrophic or mesotrophic habitat.

Osmosis (*ozz-MOE-sis*): A random physical process by which water molecules tend to move into areas with relatively high concentrations of salts or other dissolved substances.

Overturn: A process within a lake by which seasonal or weather-driven changes in the density of water cause the upper and lower layers of the lake to mix. In the temperate zones, seasonal mixing usually occurs during spring and autumn.

Paleozoic (*pay-lee-oe-ZOE-ik*): A period of geologic time between 600 and 250 million years ago.

pH (*pee-AYCH*): A measure of acidity related to the activity of hydrogen ions in a solution, in which each whole step along the pH scale represents a tenfold change in acidity. The more hydrogen ion activity, the more acidity but the lower the pH value.

Plankton: Organisms who are too small to resist the motion of surrounding waters and therefore tend to drift with currents. In that context, small photosynthetic algae and cyanobacteria are referred to as "phytoplankton," and small animals and protozoans are referred to as "zooplankton."

Pond: See "lake."

Prokaryote (*pro-KAER-ee-oat*): A bacterium or close relative of bacteria (Archaea).

Reclamation: The poisoning of a lake in order to eliminate unwanted species of fish and replace them with more desirable fish.

Rotenone: A pesticide derived from a South American plant, commonly used as a garden insecticide and as a fish-killing tool in lakes and rivers.

Secchi disc (*SEK-ee disc*): A white or black-and-white disc that is lowered into a water body in order to measure the clarity of the water. Invented by the nineteenth-century Italian scientist Angelo Secchi.

Seiche (*SAYSH*): A sloshing, wave-like disturbance within a lake, either on the surface ("surface seiche") or below the surface where two internal water layers of different density meet ("internal seiche"). Usually caused by strong winds.

Sexual selection: An evolutionary process by which organisms devote resources to attracting a mate and producing offspring, thereby becoming more likely to pass their genetic traits on to future generations even though their behavior or adaptations might otherwise seem harmful or wasteful. Examples include courtship calls, colorful plumage, and fighting over potential mates.

Stratification: The layering of water in a lake, usually from density differences due to differential heating of upper and lower layers ("thermal stratification"), or by layered differences in salinity or other dissolved substances ("chemical stratification").

Taiga (*TAI-gah*): Northern temperate forest dominated by conifers such as spruce, fir, larch, and pine.

Tectonic (tek-TAHN-ik): Having to do with large Earth movements such as faulting, seafloor spreading, or continental drift.

Trophic cascade (TROE-fik kass-KAYD): Changes in an ecosystem that may be relayed down the food chain from predator to prey ("top-down cascade") or up the food chain from nutrients to plants to grazers to predators ("bottom-up cascade").

Varve: A layer of sediment in a lake or ocean that represents a single year of deposition, usually including paired seasonal bands with differing colors.

ACKNOWLEDGMENTS

THIS BOOK IS dedicated to three remarkable men who passed away while it was being written in 2016. It has therefore unexpectedly become a memorial to them, a celebration of three long lives well lived and of the natural world that they helped me to better understand and appreciate.

Dan Livingstone, my graduate advisor at Duke University, launched many scientific careers, including my own. The lake research that he pioneered continues to inform, and the love of language that he exuded continues to inspire. Seven months after his passing in March, his former graduate student Joe Richardson died. You will have met Joe as well as Dan in the chapter on African lakes, in which I describe how they narrowly escaped being killed by a crocodile while studying a lake in Zambia. Joe was a top-flight scientist, educator, and fellow lover of Africa who supported and encouraged me throughout my career. When my father, Jay Stager, also died unexpectedly in December, I lost one of my closest friends, a cherished advisor, and a lifelong supporter of my fascination with lakes, as the above photo he took of me as a child attests. I'm grateful to have had each of them in my life for as long as I did.

This book also reflects the influence of many talented people who helped me to become a better science communicator, including Tom Canby, Rick Gore, Chris Johns, and Tony Suau at *National Geographic Magazine*; Chris Conway at the *New York Times*; Martha Foley, Joel Hurd, Brian Mann, and Ellen Rocco at North Country Public Radio; Betsy Folwell, Christopher Shaw, Annie Stoltie, and Mary Thill at *Adirondack Life Magazine*; Rob Carr and Stephanie Ratcliffe at The Wild Center; and Maurice Kenny and Nathalie Thill at the Adirondack Center for Writing. My editor at W. W. Norton, John Glusman, my to-die-for agent Sandy Dijkstra, and her staff at Sandra Dijkstra Literary Agency guided this project with great skill and patience on its journey into print. My mother, Asha Stager, and my father's wife Devora Stager also edited early drafts of the book carefully and wisely.

The Concord Museum in Concord, Massachusetts, provided the images of Thoreau's map of Walden Pond and William Stillman's painting of the Philosophers' Camp, Nancy Bernstein produced the beautiful hand-drawn maps, and Kary Johnson, Bob Kendall, and Brendan Wiltse kindly provided additional photos. Paul Smith's College student Hanna Cromie wrote the wonderful poem in the prologue, which is used with her permission. I would like to thank Paul Smith's College students Ken Alton, Jason Fitzgerald, Rory Fraser, Alex Garrigan-Piela, Dustin Grzesik, Scott Haddam, Matt Hazzard, Carlene Heimiller, Elliott Lewis, Jerome Madsen, Josh Paradis, David Prosser, Kristen Przywara, Sean Regalado, Matt Spadoni, and Ben Wrazen, who provided assistance in the field and lab as well as fresh insights and inspiration. Additional assistance came from Ndoni Sangwa Paul in Cameroon; Kevin Watkins in East Africa; Brad Hubeny, Jennifer Ingram, Jacqui Kluft, and Richard Primack at Walden Pond; and Colin Beier, Dirk Bryant, Stuart Buchanan, Mike Carr, Charlotte Demers, John Fadden, Ed Hixson, Tom Lake, Ted Mack, Brian McDonnell, Rich Preall, Jay Swartz, Mary Thill, and Mark Wilson in the Adirondacks.

Thanks also go to Gary Roper, who gave me my first microscope and thereby opened the world of the Lily Pond to me. Many other scientist-colleagues helped to shape the experiences and ideas described on these pages as well, including Kristina Arseneau, Colin Beier, Larry Cahoon,

Doug Carlson, Brian Chase, Jim Coffman, Andy Cohen, Brian Cumming, Francoise Gasse, John Glew, Moshe Gophen, Kurt Haberyan, Tom Holsen, Tom Johnson, Dan Kelting, Bob Kendall, George Kling, Kate Laird, Corey Laxson, Melissa Lenker, Jon Lothrop, Jean Maley, Michael Martin, Rodney Maud, Paul Mayewski, Mike Meadows, Tim Messner, Craig Milewski, Brenda Miskimmin, Pat Palmer, Joe Richardson, Karen Roy, Thom Sanger, Chris Scholz, John Smol, Lee Ann Sporn, Brendan Wiltse, and Susan Winchell-Sweeney. My research on lakes has been funded by the National Science Foundation, the Sky Mountain Fund, the A. C. Walker Foundation, and the Draper-Lussi Endowed Chair position at Paul Smith's College through the support of people like Ray Agnew, Caroline and Sergei Lussi, John Mills, Paul and Nancy Soderholm, and Dave Verardo.

Throughout this project, and especially during the triumphs and tragedies of 2016, my wife, Kary Johnson, has been my deepest and most reliable source of ideas, enthusiasm, and encouragement. She has cheerfully traveled with me to lakes near and far under all kinds of conditions, and I am pleased and proud to share some of her remarkable spirit and beautiful photographs with you on these pages.

REFERENCES CITED

PROLOGUE

Dillard, A. 1974. *Pilgrim at Tinker Creek*. New York: Harper Collins.

Hutchinson, G. E. 1957. *A Treatise on Limnology. Volume I. Geography, Physics, and Chemistry*. New York: John Wiley.

———. 1967. *A Treatise on Limnology. Volume II. Introduction to Lake Biology and the Limnoplankton*. New York: John Wiley and Sons.

———. 1975. *A Treatise on Limnology. Volume III. Limnological Botany*. New York: John Wiley and Sons.

Hutchinson, G. E., and Y. H. Edmondson. 1993. *A Treatise on Limnology. Volume IV. The Zoobenthos*. New York: John Wiley and Sons.

Kimmerer, R. W. 2013. *Braiding Sweetgrass: Indigenous Wisdom, Scientific Knowledge and the Teachings of Plants*. Minneapolis: Milkweed Editions.

Stager, C. 2011. *Deep Future: The Next 100,000 Years of Life on Earth*. New York: St. Martin's Press.

———. 2014. *Your Atomic Self: The Invisible Elements That Connect You to Everything Else in the Universe*. New York: Saint Martin's Press.

Thoreau, H. D. 1854. *Walden; or Life in the Woods*. Boston: Ticknor and Fields.

Wagner, M. 2015. Do it for love. *Science* 348: 1394.

Wetzel, R. G. 2001. *Limnology: Lake and River Ecosystems*. 3rd edition. London: Academic Press.

CHAPTER ONE: WALDEN

Anselmetti, F. S., D. A. Hodell, D. Ariztegui, M. Brenner, and M. F. Rosenmeier. 2007. Quantification of soil erosion rates related to ancient Maya deforestation. *Geology* 35: 915–918.

Blanke, S., and B. Robinson. 1985. From Musketaquid to Concord: The Native and European Experience. Concord Antiquarian Museum, Concord, MA.

Bonatto, S. L., and F. M. Salzano. 1997. A single and early migration for the peopling of the Americas supported by mitochondrial DNA sequence data. *Proceedings of the National Academy of Sciences* 94: 1866–1871.

Bostock, J., and H. T. Riley. 1855. *Pliny the Elder, The Natural History*. Perseus at Tufts, New York.

Burroughs, J. 1920. *Accepting the Universe: Essays in Naturalism*. New York: Houghton Mifflin.

Colman, J. A., and P. J. Friesz. 2001. Geohydrology and limnology of Walden Pond, Concord, Massachusetts. US Geological Survey. *Water-Resources Investigations Report* 01–4137. Northborough, MA.

Deevey, E. S., Jr. 1942. A re-examination of Thoreau's "Walden." *Quarterly Review of Biology* 17: 1–11.

Doucette, D. L. 2005. Reflections of the Middle Archaic: A view from Annasnappet Pond. *Bulletin of the Massachusetts Archaeological Society* 66: 22–33.

Douglas, M. S. V., J. P. Smol, J. M. Savelle, and J. M. Blais. 2004. Prehistoric Inuit whalers affected Arctic freshwater ecosystems. *Proceedings of the National Academy of Sciences* 101: 1613–1617.

Ekdahl, E. J., J. L. Teranes, T. P. Guilderson, C. L. Turton, J. H. McAndrews, C. A. Wittkop, and E. F. Stoermer. 2004. Prehistorical record of cultural eutrophication from Crawford Lake, Canada. *Geology* 32: 745–748.

Ekdahl, E. J., J. L. Teranes, C. A. Wittkop, E. F. Stoermer, E. D. Reavie, and J. P. Smol. 2007. Diatom assemblage response to Iroquoian and Euro-Canadian eutrophication of Crawford Lake, Ontario, Canada. *Journal of Paleolimnology* 37: 233–246.

Feranec, R. S., N. G. Miller, J. C. Lothrop, and R. W. Graham. 2011. The *Sporormiella* proxy and end-Pleistocene megafaunal extinction: A perspective. *Quaternary International* 245: 333–338.

Greene, H. W. 2005. Organisms in nature as a central focus for biology. *Trends in Ecology and Evolution* 20: 23–27.

Haynes, G. 2007. A review of some attacks on the overkill hypothesis, with special attention to misrepresentations and doubletalk. *Quaternary International* 169: 84–94.

Holtgrieve, G. W., et al. 2011. A coherent signature of anthropogenic nitrogen deposition to remote watersheds of the northern hemisphere. *Science* 334: 1545–1548.

Köster, D., R. Pienitz, B. B. Wolfe, S. Barry, D. R. Foster, and S. S. Dixit. 2005. Paleolimnological assessment of human-induced impacts on Walden Pond (Massachusetts, USA) using diatoms and stable isotopes. *Aquatic Ecosystem Health and Management* 8: 117–131.

Lewis, S. L., and M. A. Maslin. 2015. Defining the Anthropocene. *Nature* 519: 171–180.

Maynard, W. B. 2004. *Walden Pond: A History*. New York: Oxford University Press.

Mayr, E. 1982. Biology is not postage stamp collecting. *Science* 216: 718–720.

McCutchen, C. W. 1970. Surface films compacted by moving water: Demarcation lines reveal film edges. *Science* 170: 61–64.

McDowell, R. S., and C. W. McCutchen. 1971. The Thoreau-Reynolds ridge, a lost and found phenomenon. *Science* 172: 973.

McLauchlan, K. K., J. J. Williams, J. M. Craine, and E. S. Jeffers. 2013. Changes in global nitrogen cycling during the Holocene epoch. *Nature* 495: 352–355.

McNeil, C. L., D. A. Burney, and L. P. Burney. 2010. Evidence disputing deforestation as the cause for the collapse of the ancient Maya polity of Copan, Honduras. *Proceedings of the National Academy of Sciences* 107: 1017–1022.

Munoz, S. E., K. Gajewski, and M. C. Peros. 2010. Synchronous environmental and cultural change in the prehistory of the northeastern United States. *Proceedings of the National Academy of Sciences* 107: 22008–22013.

Primack, R. B. 2014. *Walden Warming: Climate Change Comes to Thoreau's Woods*. Chicago: University of Chicago Press.

Ruddiman, W. F., E. C. Ellis, J. O. Kaplan, and D. Q. Fuller. 2015. Defining the epoch we live in. *Science* 348: 38–39.

Rühland, K. M., A. M. Paterson, and J. P. Smol. 2015. Lake diatom responses to warming: reviewing the evidence. *Journal of Paleolimnology* 54, doi: 10.1007/s10933-015-9837-3.

Sandom, C., S. Faurby, B. Sandel, and J.-C. Svenning. 2014. Global late Quaternary megafauna extinctions linked to humans, not climate. *Proceedings of the Royal Society B* 281: 20133254.

Schindler, D. W. 2012. The dilemma of controlling cultural eutrophication of lakes. *Proceedings of the Royal Society B*, doi: 10.1098/rspb.2012.1032.

Shattuck, L. 1835. *History of the Town of Concord, Middlesex County, Massachusetts, From Its Earliest Settlement to 1832*. Boston: Russell, Odiorne.

Shuman, B., J. Bravo, J. Kaye, J. A. Lynch, P. Newby, and T. Webb III. Late-Quaternary water-level variations and vegetation history at Crooked Pond, southeastern Massachusetts. 2001. *Quaternary Research* 56: 401–410.

Sivarajah, B., K. M. Rühland, A. L. Labaj, A. M. Paterson, and J. P. Smol. 2016. Why is the relative abundance of *Asterionella formosa* increasing in a Boreal Shield lake as nutrient levels decline? *Journal of Paleolimnology* 55: 357–367.

Thackeray, S. J., I. D. Jones, and S. C. Maberly. 2008. Long-term change in the phenology of spring phytoplankton: Species-specific responses to nutrient enrichment and climatic change. *Journal of Ecology* 96: 523–535.

Thoreau, H. D. 1854. *Walden; or Life in the Woods*. Boston: Ticknor and Fields.

Waters, C. N., et al. 2016. The Anthropocene is functionally and stratigraphically distinct from the Holocene. *Science* 351: 137.

Willis, K. J., and H. J. B. Birks. 2006. What is natural? The need for a long-term perspective in biodiversity conservation. *Science* 314: 1261–1265.

Winkler, M. G. 1993. Changes at Walden Pond during the last 600 years. In E. A. Schofield and R. C. Baron (eds.), *Thoreau's World and Ours: A Natural Legacy*, 199–211. Golden, CO: North American Press.

Wolfe, A. P., and B. B. Perren. 2001. Chrysophyte microfossils record marked responses to recent environmental changes in high- and mid-arctic lakes. *Canadian Journal of Botany* 79: 747–752.

Wolfe, A. P., et al. 2013. Stratigraphic expressions of the Holocene-Anthropocene transition revealed in sediments from remote lakes. *Earth-Science Reviews* 1: 17–34.

CHAPTER TWO: WATERS OF LIFE, WATERS OF DEATH

Browne, D. R., and J. B. Rasmussen. 2009. Shifts in the trophic ecology of brook trout resulting from interactions with yellow perch: an intraguild predator-prey interaction. *Transactions of the American Fisheries Society* 138: 1109–1122.

Burroughs, J. 1871. *Wake-Robin*. New York: William H. Wise.

Cannings, S., and R. Cannings. 2014. *The New B.C. Roadside Naturalist: A Guide to Nature along B.C. Highways*. Vancouver: Greystone Books.

Carlson, D. M., and R. A. Daniels. 2004. Status of fishes in New York: Increases, declines, and homogenization of watersheds. *American Midland Naturalist* 152: 104–139.

Carpenter, S. R., J. F. Kitchell, and J. R. Hodgson. 1985. Cascading trophic interactions and lake productivity. *BioScience* 35: 634–639.

Demong, L. 2001. The use of rotenone to restore native brook trout in the Adirondack Mountains of New York—An overview. In Cailteux, R. L., et al. (eds.), *Rotenone in fisheries: Are the rewards worth the risks?*, 29–35. American Fisheries Society, Trends in Fisheries Science and Management 1, Bethesda, MD.

George, C. J. 1981. The fishes of the Adirondack Park. Publications Bulletin FW-P171, New York State Department of Environmental Conservation.

Grabar, H. 2013. 50 years after its discovery, acid rain has lessons for climate change. *Atlantic Citylab*, September 10, 2013. Accessed August 22, 2016. http://www.citylab.com/tech/2013/09/50-years-after-its-discovery-acid-rain-offers-lesson-climate-change/6837/.

Harig, A. L., and M. B. Bain. 1995. Restoring the indigenous fishes and biological integrity of Adirondack mountain lakes. A research and demonstration project in restoration ecology. New York Cooperative

Fish and Wildlife Research Unit, Department of Natural Resources, Cornell University, Ithaca, NY.

Johnson, W. D., G. F. Lee, and D. Spyridakis. 1966. Persistence of toxaphene in treated lakes. *International Journal of Air and Water Pollution* 10: 555–560.

Josephson, D. C., J. M. Robinson, J. Chiotti, and C. E. Kraft. 2014. Chemical and biological recovery from acid deposition within the Honnedaga Lake watershed, New York, USA. *Environmental Monitoring and Assessment* 186: 4391–4409.

Krieger, D. A., J. W. Terrell, and P. C. Nelson. 1983. Habitat suitability information: Yellow perch. US Fish and Wildlife Service, FWS/OBS-82/10.55.

Lennon, R. E. 1970. Control of freshwater fish with chemicals. *Proceedings of the 4th Vertebrate Pest Conference.* Paper 25. http://digitalcommons.unl.edu/vpcfour/25.

Leopold, A. 1949. *A Sand County Almanac.* New York: Ballantine Books.

Mather, F. 1884. Memoranda relating to Adirondack fishes with descriptions of new species from researches made in 1882. New York State Land Survey, Appendix E, 113–182.

Miskimmin, B., and D. W. Schindler. 1994. Long-term invertebrate community response to toxaphene treatment in two lakes: 50-yr records reconstructed from lake sediments. *Canadian Journal of Fisheries and Aquatic Sciences* 51: 923–932.

Miskimmin, B., P. R. Leavitt, and D. W. Schindler. 1995. Fossil record of cladoceran and algal responses to fishery management practices. *Freshwater Biology* 34: 177–190.

Mitchell, M. J., C. T. Driscoll, P. J. McHale, K. M. Roy, and Z. Dong. 2013. Lake/watershed sulfur budgets and their response to decreases in atmospheric sulfur deposition: Watershed and climate controls. *Hydrological Processes* 27: 710–720.

Muir, J. 1911. *My First Summer in the Sierra.* Boston: Houghton Mifflin.

Munro, C. L., and J. L. MacMillan. 2012. Growth and overpopulation of yellow perch and the apparent effect of increased competition on brook trout in Long Lake, Halifax, Nova Scotia. *Proceedings of the Nova Scotian Institute of Science* 47: 131–141.

Nakano, S., and M. Murakami. 2001. Reciprocal subsidies: Dynamic interdependence between terrestrial and aquatic food webs. *Proceedings of the National Academy of Sciences* 98: 166–170.

Oliver, R. L., and A. E. Walsby. 1984. Direct evidence for the role of light-mediated gas vesicle collapse in the buoyancy regulation of Anabaena flos-aquae (cyanobacteria). *Limnology and Oceanography* 29: 879–886.

Post, J. R., M. Sullivan, S. Cox, N. P. Lester, C. J. Walters, E. A. Parkinson, A. J. Paul, L. Jackson, and B. J. Shuter. 2002. Canada's recreational fisheries: The invisible collapse? *Fisheries* 27: 6–17.

Reynolds, C. S. , R. L. Oliver, and A. E. Walsby. 1987. Cyanobacterial dominance: The role of buoyancy regulation in dynamic lake environments. *New Zealand Journal of Marine and Freshwater Research* 21: 379–390.

Sepulveda-Villet, O. J., A. M. Ford, J. D. Williams, and C. A. Stepien. 2009. Population genetic diversity and phylogeographic divergence patterns of the yellow perch (*Perca flavescens*). *Journal of Great Lakes Research* 35: 107–119.

Sepulveda-Villet, O. J., and C. A. Stepien. 2012. Waterscape genetics of the yellow perch (*Perca flavescens*): Patterns across large connected ecosystems and isolated relict populations. *Molecular Ecology* 21: 5795–5826.

Stager, J. C. 2001. Did reclamation pollute Black Pond? *Adirondack Journal of Environmental Studies* 8: 22–27.

Stager, J. C., P. R. Leavitt, and S. Dixit. 1997. Assessing impacts of past human activity on water quality in Upper Saranac Lake, NY. *Lake and Reservoir Management* 13: 175–184.

Stager, J. C., L. A. Sporn, M. Johnson, and S. Regalado. 2015. Of paleogenes and perch: What if an "alien" is actually a native? *PLOS ONE*, doi: 10.1371/journal.pone.0119071.

Vadeboncoeur, Y., E. Jeppesen, M. J. Vander Zanden, H.-H. Schierup, K. Christoffersen, and D. M. Lodge. 2003. From Greenland to green lakes: Cultural eutrophication and the loss of benthic pathways in lakes. *Limnology and Oceanography* 48: 1408–1418.

Winslow, R. 2015. The one about the invading perch turns out to be a fish tale. *Wall Street Journal*, July 28, 2015. Accessed August 19, 2016. http://www.wsj.com/articles/the-one-about-the-invading-perch-turns-out-to-be-a-fish-tale-1438123309.

CHAPTER THREE: LAKES THROUGH THE LOOKING GLASS

Cael, B. B., and D. A. Seekel. 2016. The size-distribution of Earth's lakes. *Nature Scientific Reports*, doi: 10.1038/srep29633.

Cooper, E. K. 1960. *Science on the shores and banks*. New York: Harcourt, Brace, and World.

Franssen, J. J. H., and S. C. Scherrer. 2008. Freezing of lakes on the Swiss plateau in the period 1901–2006. *International Journal of Climatology* 28: 421–433.

Kling, G. W. 1987. Comparative limnology of lakes in Cameroon, West Africa. PhD diss., Duke University.

Klots, E. B. 1966. *The New Field Guide of Freshwater Life*. New York: G. P. Putnam's Sons.

Loizeau, J.-L., and J. Dominik. 2005. The history of eutrophication and restoration of Lake Geneva. *Terre et Environnement* 50: 43–56.

Miller, P. L. 1987. *Dragonflies*. Cambridge: Cambridge University Press.

Panneton, W. M. 2013. The mammalian dive response: An enigmatic reflex to preserve life? *Physiology* 28: 284–297.

Thorpe, S. A. 1971. Asymmetry of the internal seiche in Loch Ness. *Nature* 231: 306–308.

———. 1972. The internal surge in Loch Ness. *Nature* 237: 96–98.

Vincent, W. F., and C. Bertola. 2012. François Alphonse Forel and the oceanography of lakes. *Archives des Sciences* 65: 51–64.

Wedderburn, E. M. 1904. Seiches observed in Loch Ness. *Geographical Journal* 24: 441–442.

Wetzel, R. G. 2001. *Limnology: Lake and River Ecosystems*. 3rd edition. London: Academic Press.

Yen, J. 2000. Life in transition: Balancing inertial and viscous forces by planktonic copepods. *Biological Bulletin* 198: 213–224.

Yen, J., J. K. Sehn, K. Catton, A. Kramer, and O. Sarnelle. 2011. Pheromone trail following in three dimensions by the freshwater copepod *Hesperodiaptomus shoshone*. *Journal of Plankton Research* 33: 907–916.

CHAPTER FOUR: THE GREAT RIFT

Allendorf, F. W., and J. J. Hard. 2009. Human-induced evolution caused by unnatural selection through harvest of wild animals. *Proceedings of the National Academy of Sciences* 106: 9987–9994.

Alós, J., M. Palmer, and R. Arlinghaus. 2012. Consistent selection towards low activity phenotypes when catchability depends on encounters among human predators and fish. *PLOS ONE* 7: e48030, doi:10.1371/journal.pone.0048030.

Balirwa, J. S., et al. 2003. Biodiversity and fishery sustainability in the Lake Victoria basin: An unexpected marriage? *BioScience* 53: 703–715.

Beadle, L. C. 1981. *The inland waters of tropical Africa*. London: Longman.

Beuving, J. J. 2010. Playing pool along the shores of Lake Victoria: Fishermen, careers and capital accumulation in the Ugandan Nile perch business. *Africa* 80: 224–248.

Darwall, W. R. T., E. H. Allison, G. F. Turner, and K. Irvine. 2010. Lake of flies, or lake of fish? A trophic model for Lake Malawi. *Ecological Modeling* 22: 713–727.

Darwin, C. 1859. *On the Origin of Species*. London: John Murray.

Downing, A. S., et al. 2014. Coupled human and natural system dynamics as key to the sustainability of Lake Victoria's ecosystem services. *Ecology and Society* 19: 31. http://dx.doi.org/10.5751/ES-06965–190431.

Eccles, D. H. 1974. An outline of the physical limnology of Lake Malawi (Lake Nyasa). *Limnology and Oceanography* 19: 730–742.

Eschenbach, W. W. 2004. Climate-change effect on Lake Tanganyika? *Nature* 430, doi:10.1038/nature02689.

Geheb, K. 1999. Small-scale regulatory institutions in Kenya's Lake Victoria fishery. In Kawanabe, H., G. W. Coulter, and A. C. Roosevelt (eds.), *Ancient Lakes: Their Cultural and Biological Diversity*, 113–121. Ghent: Kenobi Productions.

Huckabay, J. D. 1983. The fisheries of northern Zambia. In Ooi, J.-B. (ed.), *Natural Resources in Tropical Countries*, chapter 7. Singapore: Singapore University Press.

Kendall, R. L. 1969. An ecological history of the Lake Victoria basin. *Ecological Monographs* 39: 121–176.

Lewin, R. 1981. Lake bottoms linked with human origins. *Science* 211: 564–566.

Maan, M. E., Seehausen, O., and J. J. M. Van Alphen. 2010. Female preferences and male coloration covary with water transparency in a Lake Victoria cichlid fish. *Biological Journal of the Linnean Society* 99: 398–406.

O'Reilly, C. M., S. R. Alin, P.-D. Plisnier, A. S. Cohen, and B. A. McKee. 2003. Climate change decreases aquatic ecosystem productivity of Lake Tanganyika, East Africa. *Nature* 424: 766–768.

Pandolfi, J. M. 2009. Evolutionary impacts of fishing: Overfishing's 'Darwinian debt.' *F1000 Biology Reports* 1: 43–45.

Post, J. R., M. Sullivan, S. Cox, N. P. Lester, C. J. Walters, E. A. Parkinson, A. J. Paul, L. Jackson, and B. J. Shuter. 2002. Canada's recreational fisheries: The invisible collapse? *Fisheries* 27: 6–17.

Pringle, R. M. 2005. The origins of the Nile perch in Lake Victoria. *BioScience* 55: 780–787.

Reynolds, J. E. and D. F. Greboval. 1989. Socio-economic effects of the evolution of Nile perch fisheries in Lake Victoria: A review. Food and Agriculture Organization of the United Nations, CIFA Technical Paper 17.

Richardson, J., and D. Livingstone. 1962. An attack by a Nile crocodile on a small boat. *Copeia* 1: 203–204.

Rosendahl, B. R., and D. A. Livingstone. 1983. Rift lakes of East Africa: New seismic data and implications for future research. *Episodes* 1: 14–19.

Salzburger, W., T. Mack, E. Verheyen, and A. Meyer. 2005. Out of Tanganyika: Genesis, explosive speciation, key-innovations and phylogeography of the haplochromine fishes. *BMC Evolutionary Biology* 5, doi: 10.1186/1471-2148-5-17.

Sarrazin, F., and J. Lecomte. 2016. Evolution in the Anthropocene. *Science* 351: 922–923.

Scholz, C. A., et al. 2007. East African megadroughts between 135 and 75 thousand years ago and bearing on early-modern human origins. *Proceedings of the National Academy of Sciences* 104: 16416–16421.

Seehausen, O. 2015. Beauty varies with the light. *Nature* 521: 34–35.

Seehausen, O., E. Koetsier, M. V. Schneider, L. J. Chapman, C. A. Chapman, M. E. Knight, G. F. Turner, J. M. van Alphen, and R. Bills. 2002. Nuclear markers reveal unexpected genetic variation and a Congolese-

Nilotic origin of the Lake Victoria cichlid species flock. *Proceedings of the Royal Society of London B* 270: 129–137.

Seehausen, O., J. J. M. van Alphen, and F. Witte. 1997. Cichlid fish diversity threatened by eutrophication that curbs sexual selection. *Science* 277: 1808–1811.

Spinney, L. 2010. Dreampond revisited. *Nature* 466: 174–175.

Stager, J. C., R. E. Hecky, D. Grzesik, B. F. Cumming, and H. Kling. 2009. Diatom evidence for the timing and causes of eutrophication in Lake Victoria, East Africa. *Hydrobiologia* 636: 463–478.

Stager, J. C., and T. C. Johnson. 2007. The late Pleistocene desiccation of Lake Victoria and the origin of its endemic biota. *Hydrobiologia* 596, doi:10.1007/210750–007–9158–2.

Stager, J. C., D. R. Ryves, B. M. Chase, and F. S. R. Pausata. 2011. Catastrophic drought in the Afro-Asian monsoon regions during Heinrich Event 1. *Science* 331: 1299–1302.

Sutter, D. A., C. D. Suski, D. P. Phillipp, T. Klefoth, D. H. Wahl, P. Kersten, S. J. Cooke, and R. Arlinghaus. 2012. Recreational fishing selectively captures individuals with the highest fitness potential. *Proceedings of the National Academy of Sciences* 109: 20960–20965.

Van Valen, L. 1973. A new evolutionary law. *Evolutionary Theory* 1: 1–30.

Vonlanthen, P., D. Bittner, A. G. Hudson, K. A. Young, R. Müller, B. Lundsgaard-Hansen, D. Roy, S. Di Piazza, C. R. Largiader, and O. Seehausen. 2012. Eutrophication causes speciation reversal in whitefish adaptive radiations. *Nature* 482: 357–362.

Wallisch, P. 2016. Unleashing the beast within. *Science* 351: 232.

Weiner, J. 2005. Evolution in action. *Natural History* 115 (9): 47–51.

Weston, M. 2015. Troubled waters: Why Africa's largest lake is in grave danger. *Slate*, March 27, 2015. Accessed August 22, 2016. http://www.slate.com/articles/news_and_politics/roads/2015/03/lake_victoria_is_in_grave_danger_africa_s_largest_lake_is_threatened_by.html.

Wilson, A. D. M., J. W. Brownscombe, B. Sullivan, S. Jain-Schlaepfer, and S. J. Cooke. 2015. Does angling technique selectively target fishes based on their behavioral type? *PLOS ONE* 10: e0135848.

CHAPTER FIVE: GALILEE

Bartov, Y., M. Stein, Y. Enzel, A. Agnon, and Z. Reches. 2002. Lake levels and sequence stratigraphy of Lake Lisan, the Late Pleistocene precursor of the Dead Sea. *Quaternary Research* 57: 9–21.

Baruch, U. 1986. The Late Holocene vegetational history of Lake Kinneret (Sea of Galilee), Israel. *Paléorient* 12: 37–48.

Baruch, U., and S. Bottema. 1999. A new pollen diagram from Lake Hula: vegetational, climatic and anthropogenic implications. In H. Kawanabe,

G. W. Coulter, and A. C. Roosevelt (eds.), *Ancient lakes: Their cultural and biological diversity*, 75–86. Ghent: Kenobi Productions.

Bar-Yosef, M. D. E., B. Vandermeersch, and O. Bar-Yosef. 2009. Shells and ochre in Middle Paleolithic Qafzeh Cave, Israel: Indications for modern behavior. *Journal of Human Evolution* 56: 307–314.

Ben David, A. 2012. Once pristine, now polluted, there's hope yet for Jordan River. *Al-Monitor* (posted October 22, 2012). Accessed August 19, 2016. http://www.al-monitor.com/pulse/culture/2012/10/saving-the-jordan .html#.

Bergoglio, J. M. (Pope Francis). 2015. *Laudato Si. On Care for Our Common Home*. Encyclical Letter of Pope Francis. Accessed August 19, 2016. http://w2.vatican.va/content/francesco/en/encyclicals/documents/ papa-francesco_20150524_enciclica-laudato-si.html.

Bocquentin, F., and O. Bar-Yosef. 2004. Early Natufian remains: evidence for physical conflict from Mt. Carmel, Israel. *Journal of Human Evolution* 47: 19–23.

Bodaker, I., O. Béjà, I. Sharon, R. Feingersch, M. Rosenberg, A. Oren, M. Y. Hindiyeh, and H. I. Malkawi. 2009. Archaeal diversity in the Dead Sea: Microbial survival under increasingly harsh conditions. *Natural Resources and Environmental Issues* 15, Article 25. http:// digitalcommons.usu.edu/nrei/vol15/iss1/25.

Bowles, S. 2011. Cultivation of cereals by the first farmers was not more productive than foraging. *Proceedings of the National Academy of Sciences* 108: 4760–4765.

Bowles, S., and J. K. Choi. 2013. Coevolution of farming and private property during the early Holocene. *Proceedings of the National Academy of Sciences* 110: 8830–8835.

Campbell, J. 1988. *The Power of Myth*. New York: Doubleday.

Clarke, J., et al. 2016. Climatic changes and social transformations in the Near East and North Africa during the 'long' 4th millennium BC: A comparative study of environmental and archaeological evidence. *Quaternary Science Reviews* 136: 96–121.

Cole, T. 1836. Essay on American scenery. *American Monthly Magazine* 1 (January): 1–12.

Cordova, C. E. 2007. *Millennial Landscape Change in Jordan: Geoarchaeology and Cultural Ecology*. Tucson: University of Arizona Press.

Cronon, W. 1995. The trouble with wilderness. In William Cronon (ed.), *Uncommon Ground: Rethinking the Human Place in Nature*, 69–90. New York: W. W. Norton.

Degani, G., Y. Yehuda, K. Jackson, and M. Gophen. 1998. Temporal variation in fish community structure in a newly created wetland lake (Lake Agmon) in Israel. *Wetland Ecology and Management* 6: 151–157.

Dell'Amore, C. 2011. New life-forms found at bottom of Dead Sea. *National Geographic News*, September 30, 2011. Accessed August 19, 2016. http://

news.nationalgeographic.com/news/2011/09/110928-new-life-dead-sea-bacteria-underwater-craters-science/.

Doebley, J. 2006. Unfallen grains: How ancient farmers turned weeds into crops. *Science* 312: 1318–1319.

Drake, N. A., R. M. Blench, S. J. Armitage, C. S. Bristow, and K. H. White. 2011. Ancient watercourses and biogeography of the Sahara explain the peopling of the desert. *Proceedings of the National Academy of Sciences* 108: 458–462.

Eckert, W., K. D. Hambright, Y. Z. Yacobi, I. Ostrovsky, and A. Sukenik. 2002. Internal wave induced changes in the chemical stratification in relation to the thermal structure in Lake Kinneret. *Verhandlungen des Internationalen Verein Limnologie* 28: 962–966.

Frumkin, A., M. Margaritz, I. Carmi, and I. Zak. 1991. The Holocene climatic record of the salt caves of Mount Sedom, Israel. *Holocene* 1: 191–200.

Gaudzinski, S. 2004. Subsistence patterns of Early Pleistocene hominids in the Levant—taphonomic evidence from the Ubeidiya Formation (Israel). *Journal of Archaeological Science* 31: 65–75.

Gibbons, A. 2016. First farmers' motley roots. *Science* 353: 207–208.

Gophen, M. 1986. Fisheries management in Lake Kinneret (Israel). *Lake and Reservoir Management* 2: 327–332.

———. 1989. Utilization and water quality management of Lake Kinneret, Israel. *Toxicity Assessment* 4: 353–362.

———. 2000. Lake Kinneret (Israel) ecosystem: Long-term instability or resiliency? *Water, Air, and Soil Pollution* 123: 323–335.

Goren-Inbar, N., C. S. Feibel, K. L. Verosub, Y. Melamed, M. E. Kislev, E. Tchernov, and I. Saragusti. 2000. Pleistocene milestones on the Out-Of-Africa corridor at Gesher Benot Ya'aqov, Israel. *Science* 289: 944–947.

Grosman, L., N. D. Munro, and A. Belfer-Cohen. 2008. A 12,000-year-old shaman burial from the southern Levant (Israel). *Proceedings of the National Academy of Sciences* 105: 17665–17669.

Grosman, L., N. D. Munro, I. Abadi, E. Boaretto, D. Shaham, A. Belfer-Cohen, and O. Bar-Yosef. 2016. Nahal Ein Gev II, a Late Natufian community at the Sea of Galilee. *PLOS ONE* 11: e0146647.

Hambright, K. D., S. C. Blumenshine, and J. Shapiro. 2002. Can filter-feeding fishes improve water quality in lakes? *Freshwater Biology* 47: 1–10.

Hambright, K. D., and T. Zohary. 1998. Lakes Hula and Agmon: Destruction and creation of wetland ecosystems in northern Israel. *Wetlands Ecology and Management* 6: 83–89.

Hammer, M. F., et al. 2000. Jewish and Middle Eastern non-Jewish populations share a common pool of Y-chromosome biallelic haplotypes. *Proceedings of the National Academy of Sciences* 97: 6769–6774.

Harari, Y. N. 2014. Were we happier in the stone age? *Guardian*, September 5, 2014. Accessed August 22, 2016. https://www.theguardian.com/books/2014/sep/05/were-we-happier-in-the-stone-age.

Hazan, N., M. Stein, A. Agnon, S. Marco, D. Nadel, J. F. W. Negendank, M. J. Schwab, and D. Neev. 2005. The late Quaternary limnological history of Lake Kinneret (Sea of Galilee), Israel. *Quaternary Research* 63: 60–77.

Hurwitz, S., Z. Garfunkel, Y. Ben-Gai, M. Reznikov, Y. Rotstein, and H. Gvirtzman. 2002. The tectonic framework of a complex pull-apart basin: Seismic reflection observations in the Sea of Galilee, Dead Sea transform. *Tectonophysics* 359: 289–306.

Johnson, D. D. P. 2016. Hand of the gods in human civilization. *Nature* 530: 285–286.

Klinger, Y., J. P. Avouac, N. Abou Karaki, L. Dorbath, D. Bourles, and J. L. Reyss. 2000. Slip rate on the Dead Sea transform fault in northern Araba valley (Jordan). *Geophysical Journal International* 142: 755–768.

Langgut, D., F. H. Neumann, M. Stein, A. Wagner, E. J. Kagan, E. Boaretto, and I. Finkelstein. 2014. Dead Sea pollen record and history of human activity in the Judean Highlands (Israel) from the Intermediate Bronze into the Iron Ages (2500–500 BCE). *Palynology*, doi: 10.1080/01916122.2014.906001.

Leroy, S. A. G. 2010. Pollen analysis of core DS7–1SC (Dead Sea) showing intertwined effects of climatic change and human activities in the Late Holocene. *Journal of Archaeological Science* 37: 306–316.

Levi-Yadun, S., A. Gopher, and S. Abbo. 2006. How and when was wild wheat domesticated? *Science* 313: 296.

Machlis, G., and M. McNutt. 2015. Parks for science. *Science* 348: 1291.

Maher, L. A., T. Richter, and J. T. Stock. 2012. The pre-Natufian Epipaleolithic: Long-term behavioral trends in the Levant. *Evolutionary Anthropology* 21: 69–81.

Manning, K., and A. Timpson. 2014. The demographic response to Holocene climate change in the Sahara. *Quaternary Science Reviews* 101: 28–35.

McDermott, F., R. Grun, C. B. Stringer, and C. J. Hawkesworth. 1993. Mass-spectrometric U-series dates for Israeli Neanderthal/early modern hominid sites. *Nature* 363: 252–255.

Migowski, C., M. Stein, S. Prasad, J. F. W. Negendank, and A. Agnon. 2006. Holocene climate variability and cultural evolution in the Near East from the Dead Sea sedimentary record. *Quaternary Research* 66: 421–431.

Munro, N. D., and L. Grosman. 2010. Early evidence (ca. 12,000 BP) for feasting at a burial cave in Israel. *Proceedings of the National Academy of Sciences* 107: 15362–15366.

Nadel, D., E. Weiss, O. Simchoni, A. Tsatskin, A. Danin, and M. Kislev. 2004. Stone Age hut in Israel yields world's oldest evidence of bedding. *Proceedings of the National Academy of Sciences* 101: 6821–6826.

Nadel, D., et al. 2013. Earliest floral grave lining from 13,700–11,700-y-old Natufian burials at Raqefet Cave, Mt. Carmel, Israel. *Proceedings of the National Academy of Sciences* 110: 11774–11778.

Neugebauer, I. 2015. Reconstructing climate from the Dead Sea sediment records using high-resolution facies analyses. PhD diss., University of Potsdam.

Nof, D., I. McKeague, and N. Paldor. 2006. Is there a paleolimnological explanation for 'walking on water' in the Sea of Galilee? *Journal of Paleolimnology* 35: 417–439.

Ostrovsky, I., Y. Z. Yacobi, P. Walline, and I. Kalikhman. 1996. Seiche-induced mixing: Its impact on lake productivity. *Limnology and Oceanography* 41: 323–332.

Payne, R. J. 2011/2012. A longer-term perspective on human exploitation and management of peat wetlands: The Hula Valley, Israel. *Mires and Peat* 9: 1–9.

Pollingher, U. 1986. Non-siliceous algae in a 5 meter core from Lake Kinneret (Israel). *Hydrobiologia* 143: 213–216.

Pollingher, U., A. Ehrlich, and S. Serruya. 1986. The planktonic diatoms of Lake Kinneret (Israel) during the last 5000 years—their contribution to the algal biomass. In M. Ricard (ed.), *Proceedings of the 8th International Diatom Symposium, Koeltz*, 459–470. Koenigstein, Germany.

Prange, M., H. Arz, and F. Lamy. 2012. Comment on "Is there a paleolimnological explanation for 'walking on water' in the Sea of Galilee?" *Journal of Paleolimnology* 38: 589–593.

Rambeau, C. M. C. 2010. Palaeoenvironmental reconstruction in the Southern Levant: Synthesis, challenges, recent developments and perspectives. *Philosophical Transactions of the Royal Society A* 368: 5225–5248.

Rosen, A. M., and I. Rivera-Collazo. 2012. Climate change, adaptive cycles, and the persistence of foraging economies during the late Pleistocene/Holocene transition in the Levant. *Proceedings of the National Academy of Sciences* 109: 3640–3645.

Rossignol-Strick, M. 1999. The Holocene climatic optimum and pollen records of sapropel 1 in the Eastern Mediterranean, 9000–6000 BP. *Quaternary Science Reviews* 18: 515–530.

Schiebel, V. 2013. Vegetation and climate history of the southern Levant during the last 30,000 years based on palynological investigation. PhD diss., University of Bonn.

Schuster, M., C. Roquin, P. Duringer, M. Brunet, M. Caugy, M. Fontugne, H. T. Mackaye, P. Vignaud, and J.-F. Ghienne. 2005. Holocene Lake Mega-Chad palaeoshorelines from space. *Quaternary Science Reviews* 24: 1821–1827.

Schwab, M. J., F. Neumann, T. Litt, J. F. W. Negendank, and M. Stein. Holocene palaeoecology of the Golan Heights (Near East): Investigation of lacustrine sediments from Birkat Ram crater lake. *Quaternary Science Reviews* 23: 1723–1731.

Schwartzstein, P. 2014. Biblical waters: Can the Jordan River be saved? *National Geographic News*, February 22, 2014. Accessed August 19, 2016.

http://news.nationalgeographic.com/news/2014/02/140222-jordan-river-syrian-refugees-water-environment/.
Scudellari, M. 2015. Myths that will not die. *Nature* 528: 322–325.
Sender, R., S. Fuchs, and R. Milo. 2016. Revised estimates for the number of human and bacteria cells in the body. *PLOS Biology*, doi: 10.1371/journal.pbio.1002533.
Sherwood, H. 2010. Pollution fears at River Jordan pilgrimage spot. *Guardian*, July 26, 2010. Accessed August 19, 2016. https://www.theguardian.com/world/2010/jul/26/israel-closes-jordan-christ-baptism.
Shteinman, B., W. Eckert, S. Kaganowsky, and T. Zohary. 1997. Seiche-induced resuspension in Lake Kinneret: A fluorescent tracer experiment. In *The Interactions Between Sediments and Water*, 123–131. Proceedings of the 7th International Symposium, Baveno, Italy, 22–25 September, 1996. Springer, Netherlands.
Sufian, S. M. 2007. *Healing the Land and the Nation: Malaria and the Zionist Project in Palestine*, 1920–1947. Chicago: University of Chicago Press.
Van der Steen, E. 2014. *Near Eastern Tribal Societies During the Nineteenth Century: Economy, Society and Politics between Tent and Town*. New York: Routledge.
Weinberger, R., Z. B. Begin, N. Waldmann, M. Gardosh, G. Baer, A. Frumkin, and S. Wdowinski. 2006. Quaternary rise of the Sedom diapir, Dead Sea basin. *GSA Special Papers* 401: 33–51.
Zohar, I., and R. Biton. 2011. Land, lake, and fish: Investigation of fish remains from Gesher Benot Ya'aqov (paleo-Lake Hula). *Journal of Human Evolution* 60: 343–356.

CHAPTER SIX: SKY WATER

Ballantyne, C. K., and J. O. Stone. 2011. Did large ice caps persist on low ground in north-west Scotland during the Lateglacial Interstade? *Journal of Quaternary Science* 27: 297–306.
Bennett, S., and A. J. Shine. 1993. Review of current work on Loch Ness sediments. *Scottish Naturalist* 105: 56–63.
Bhatla, A. 2011. Body water cycle linked to global water cycle? Accessed August 23, 2016. http://indigenouswater.blogspot.com/2011/08/body-water-cycle-linked-to-global-water.html.
Brusatte, S. L., et al. 2015. Ichthyosaurs from the Jurassic of Skye, Scotland. *Scottish Journal of Geology*, doi: 10.1144/sjg2014-018.
Cooper, M. C. 1998. Laminated sediments of Loch Ness, Scotland: Indicators of Holocene environmental change. PhD thesis, University of Plymouth, United Kingdom.
Ellis, W. S. 1977. Loch Ness—The lake and the legend. *National Geographic*, June, 758–779.

Fitzgerald, E. 1921. *Rubáiyát of Omar Khayyám*. 1st edition. New York: Thomas Y. Crowell.

Gier, N. F. 1986. *God, Reason, and the Evangelicals: Case Against Evangelical Rationalism*. Lanham, MD: Rowman and Littlefield.

Gleick, P. H. 1996. Water resources. In S. H. Schneider (ed.), *Encyclopedia of Climate and Weather*, 817–823. New York: Oxford University Press.

Goudzari, S. 2006. Alaskan lakes dry up. *LiveScience*, October 12. Accessed August 17, 2016. http://www.livescience.com/1093-alaskan-lakes-dry.html.

Grey, J. 2000. Tracing the elusive *Holopedium gibberum* in the plankton of Loch Ness. *Glasgow Naturalist* 23: 29–34.

Griffiths, H. I., D. S. Martin, A. J. Shine, and J. G. Evans. 1993. The ostracod fauna (Crustacea, Ostracoda) of the profundal benthos of Loch Ness. *Hydrobiologia* 254: 111–117.

Jaxybulatov, K., N. M. Shapiro, I. Koulakov, A. Mordret, M. Landès, and C. Sens-Schönfelder. 2014. A large magmatic sill complex beneath the Toba caldera. *Science* 346: 617–619.

Kintisch, E. 2015. Earth's lakes are warming faster than its air. *Science* 350: 1449.

Leeming, D. 2015. *The Handy Mythology Answer Book*. Detroit: Visible Ink Press.

Livingstone, D. A. 1954. On the orientation of lake basins. *American Journal of Science* 252: 547–554.

Martin, D., and A. Boyd. 1999. *Nessie: The Surgeon's Photograph Exposed*. London: Thorne Printing.

Montgomery, D. R. 2012. *The Rocks Don't Lie: A Geologist Investigates Noah's Flood*. New York: W. W. Norton.

Mueller, D. R., P. van Hove, D. Antoniades, M. O. Jeffries, and W. F. Vincent. 2009. High Arctic lakes as sentinel ecosystems: Cascading regime shifts in climate, ice cover, and mixing. *Limnology and Oceanography* 54: 2371–2385.

Normile, D. 2016. Nature from nurture. *Science* 351: 908–910.

O'Reilly, C., et al. 2015. Rapid and highly variable warming of lake surface waters around the globe. *Geophysical Research Letters* 42: 10773–10781.

PhysicalGeography.net. Accessed August 22, 2016. Chapter 8: Introduction to the Hydrosphere. b. The Water Cycle. http://www.physicalgeography.net/fundamentals/8b.html.

Roy-Leveillee, P., and C. R. Burn. 2010. Permafrost conditions near shorelines of oriented lakes in Old Crow Flats, Yukon Territory, 1509–1516. *GEO2010*, Canadian Geotechnical Conference, Calgary. Accessed August 17, 2016. http://pubs.aina.ucalgary.ca/cpc/CPC6–1509.pdf.

Sarangerel, O. 2000. *Riding Windhorses: A Journey Into the Heart of Mongolian Shamanism*. Rochester, VT: Destiny Books.

Scudellari, M. 2015. Myths that will not die. *Nature* 528: 322–325.

Sissons, J. B. 1979. The Loch Lomond Stadial in the British Isles. *Nature* 280: 199–202.

Smol, J. P., et al. 2005. Climate-driven regime shifts in the biological communities of arctic lakes. *Proceedings of the National Academy of Sciences* 102: 4397–4402.

Stager, C. 2014. *Your Atomic Self: The Invisible Elements That Connect You to Everything Else in the Universe*. New York: St. Martin's Press.

Thackeray, S. J., J. Grey, and R. I. Jones. 2000. Feeding selectivity of brown trout (*Salmo trutta*) in Loch Ness, Scotland. *Freshwater Forum* 13: 47–59.

Tierney, J. E., M. T. Mayes, N. Meyer, C. Johnson, P. W. Swarzenski, and J. M. Russell. 2010. Late-twentieth-century warming in Lake Tanganyika unprecedented since AD 500. *Nature Geoscience*, doi: 10.1038/NGEO865.

University of Arizona. 2005. Growth secrets of Alaska's mysterious field of lakes. *ScienceDaily*, June 28, 2015. www.sciencedaily.com/releases/2005/06/050627233623.htm.

Verpoorter, C., T. Kutser, D. A. Seekell, and L. J. Tranvik. 2014. A global inventory of lakes based on high-resolution satellite imagery. *Geophysical Research Letters* 41: 6396–6402.

Warren, D. R., J. M. Robinson, D. C. Josephson, D. R. Sheldon, and C. E. Kraft. 2012. Elevated summer temperatures delay spawning and reduce redd construction for resident brook trout (*Salvelinus fontinalis*). *Global Change Biology* 18: 1804–1811.

Watanabe, Y., and V. V. Drucker. 1999. Phytoplankton blooms in Lake Baikal, with reference to the lake's present state of eutrophication. In H. Kawanabe, G. W. Coulter, and A. C. Roosevelt (eds.), *Ancient Lakes: Their Cultural and Biological Diversity*, 217–225. Ghent: Kenobi Productions.

Wikipedia. Accessed August 23, 2016. Wind Horse. https://en.wikipedia.org/wiki/Wind_Horse.

Williamson, C. E., J. E. Saros, W. F. Vincent, and J. P. Smol. 2009. Lakes and reservoirs as sentinels, integrators, and regulators of climate change. *Limnology and Oceanography* 54: 2273–2282.

Williamson, G. R. 1988. Seals in Loch Ness. *Scientific Reports of the Whales Research Institute*, No. 39 (March 1988). Tokyo.

Willis, M. J., B. G. Herried, M. G. Bevis, and R. E. Bell. 2015. Recharge of a subglacial lake by subsurface meltwater in northeast Greenland. *Nature* 518: 223–226.

Winemiller, K. O., et al. 2016. Balancing hydropower and biodiversity in the Amazon, Congo, and Mekong. *Science* 351: 128–129.

Wu, Q., et al., 2016. Outburst flood at 1920 BCE supports historicity of China's Great Flood and the Xia dynasty. *Science* 353: 579–582.

CHAPTER SEVEN: HERITAGE LAKES

Adrian, R., et al. 2009. Lakes as sentinels of climate change. *Limnology and Oceanography* 54: 2283–2297.

Balcombe, J. 2016. Fish have feelings too. *New York Times*, May 14, 2016, Opinion Pages.

——— 2016. *What a fish knows: The Inner Lives of Our Underwater Cousins.* New York: *Scientific American*/Farrar, Straus, and Giroux.

Beier, C. M., J. C. Stella, M. Dovčiak, and S. A. McNulty. 2012. Local climatic drivers of changes in phenology at a boreal-temperate ecotone in eastern North America. *Climatic Change*, doi: 10.1007/s10584–012–0455-z.

Carlson, D., R. Morse, and E. Hekkala. 2015. Late-spawning suckers of New York's Adirondack Mountains. *American Currents* 40: 10–14.

Chandroo, K. P., I. J. H. Duncan, and R. D. Moccia. 2004. Can fish suffer? Perspectives on sentience, pain, fear and stress. *Applied Animal Behaviour Science* 86: 225–250.

Chandroo, K. P., S. Yue, and R. D. Moccia. 2004. An evaluation of current perspectives on consciousness and pain in fishes. *Fish and Fisheries* 5: 281–295.

Emerson, R.W. 1847. *Poems*. Boston: J. Munroe.

Holtgrieve, G. W., et al. 2011. A coherent signature of anthropogenic nitrogen deposition to remote watersheds of the northern hemisphere. *Science* 334: 1545–1548.

Johnson, B. M., and P. J. Martinez. 2000. Trophic economics of lake trout management in reservoirs of differing productivity. *North American Journal of Fisheries Management* 20: 127–143.

Lenker, M. A., B. C. Weidel, O. P. Jensen, and C. T. Solomon. 2016. Developing recreational harvest regulations for an unexploited lake trout population. *North American Journal of Fisheries Management* 36: 385–397.

Merton, T. 1994. *Witness to Freedom: Letters in Times of Crisis.* Selected and edited by W. H. Shannon. New York: Harcourt Brace.

Mueller, D. R., P. V. Hove, D. Antoniades, M. O. Jeffries, and W. F. Vincent. 2009. High Arctic lakes as sentinel ecosystems: Cascading regime shifts in climate, ice cover, and mixing. *Limnology and Oceanography* 54: 2371–2385.

Oelschlaeger, M. 2007. Ecological restoration, Aldo Leopold, and beauty: An evolutionary tale. *Environmental Philosophy* 4: 149–161.

Pinchot, G. 1907. The conservation of natural resources. *Outlook* 87: 291–294.

Robinson, J. M., D. C. Josephson, B. C. Weidel, and C. E. Kraft. 2010. Influence of variable summer water temperatures on brook trout growth, consumption, reproduction and mortality in an unstratified Adirondack lake. *Transactions of the American Fisheries Society* 139: 685–699.

Rühland, K. M., A. M. Paterson, and J. P. Smol. 2015. Lake diatom responses to warming: Reviewing the evidence. *Journal of Paleolimnology*, doi: 10.1007/s10933-015-9837-3.

Scheffer, M., et al. 2015. Creating a safe operating space for iconic ecosystems. *Science* 347: 1317–1319.

Schlett, J. 2015. *A Not Too Greatly Changed Eden: The Story of the Philosophers' Camp in the Adirondacks*. Ithaca, NY: Cornell University Press.

Shaw, C. 2011. Return to the source: The case for saving Follensby Pond and the Philosophers' Camp. *Adirondack Life* (July/August), 60–71.

Stager, J. C., B. F. Cumming, K. Laird, A. Garrigan-Piela, N. Pederson, B. Wiltse, C. S. Lane, J. Nester, and A. Ruzmaikin. 2016. A 1600 year record of hydroclimate variability from Wolf Lake, NY. *Holocene*, doi: 10.1177/0959683616658527.

Stager, J. C., and T. Sanger. 2003. An Adirondack "Heritage Lake." *Adirondack Journal of Environmental Studies* 10: 6–10.

Stillman, W. 1893. The Philosophers' Camp: Emerson, Agassiz, and Lowell in the Adirondacks. *Century* 46: 598–606.

Thackeray, S. J., I. D. Jones, and S. C. Maberly. 2008. Long-term change in the phenology of spring phytoplankton: Species-specific responses to nutrient enrichment and climatic change. *Journal of Ecology* 96: 523–535.

Thill, M. 2014. Lake trout and climate change in the Adirondacks: Status and long-term viability. Survey Report, Adirondack Chapter of The Nature Conservancy.

Watson, J. 2016. Bring climate change back from the future. *Nature* 534: 437.

Williamson, C. E., J. E. Saros, W. F. Vincent, and J. P. Smol. 2009. Lakes and reservoirs as sentinels, integrators, and regulators of climate change. *Limnology and Oceanography* 54: 2273–2282.

INDEX

Page numbers in *italics* refer to illustrations.